Investigating Fossils

Investigating Fossils

A History of Palaeontology

Wilson J. Wall

WILEY Blackwell

This edition first published 2021
© 2021 John Wiley & Sons Ltd

Registered Offices
John Wiley & Sons, Inc., 111 River Street, Hoboken, NJ 07030, USA
John Wiley & Sons Ltd, The Atrium, Southern Gate, Chichester, West Sussex, PO19 8SQ, UK

Editorial Office
9600 Garsington Road, Oxford, OX4 2DQ, UK

For details of our global editorial offices, customer services and more information about Wiley products, visit us at www.wiley.com.

Wiley also publishes its books in a variety of electronic formats and by print-on-demand. Some content that appears in standard print versions of this book may not be available in other formats.

Library of Congress Cataloging-in-Publication Data applied for

Paperback: 9781119698456

Cover Design: Wiley
Cover Image: © Alice Cahill/Getty Images

Set in 9.5/12.5pt STIXTwoText by Straive, Pondicherry, India

10 9 8 7 6 5 4 3 2 1

Contents

Acknowledgements

Thanks to Alison, who always encourages and helps.

I would like to praise the Society of Authors for providing generous support in the form of a travel grant.

People in the Text

Agassiz, Louis (1807–1873). Born in Switzerland, he was initially educated at home and later at more formal schools. He attended university at Zürich, Heidelberg and Munich. He received his PhD in 1829 from Erlangen and MD from Munich in 1830. In 1832, Agassiz was appointed professor of natural history at the University of Neuchâtel. He was appointed Professor of Zoology and Geology at Harvard University where he stayed until his death. In 1850, he married Elizabeth Cary, his second wife. His married his first wife, Cecilie Braun, the sister of one of his college friends, in 1833. They had three children.

Agricola, see **Bauer**

Anaximander (611–547 BCE). A Greek philosopher from Ionia, Anaximander was a pupil of Thales. Although his writings did not survive the vicissitudes of history, he is credited by third parties with many ideas new for his age. He speculated on the origins of the earth and of man. Appreciating the curvature of the horizon he imagined the earth as cylindrical and poised in space.

Anning, Mary (1799–1848). Born in Lyme Regis, Mary's father supplemented his income by searching for fossils in the coastal cliffs that could be sold. Mary carried on the tradition with notable finds, such as the *Ichthyosaur,* now in the Natural History Museum, London, in 1828 she discovered the first *Plesiosaurus,* and also the first *Pterodactyl*. She died, reportedly of breast cancer, and was buried locally. During her illness she was financially helped by the Geological Society of London.

Aristotle (384–322 BCE). Born in Stagira, northern Greece he was the son of a doctor and a member of Plato's academy. Philip of Macedonia invited Aristotle to be tutor to his son Alexander. He retired to Euboea in 323 BCE.

Bacon, Francis (1561–1626). His father was a Statesman and it was the intention that Bacon would follow the same path, starting with training in law. He became Lord High Chancellor in 1618 under James I. This success was short-lived as he was banished from Court in 1621 for taking bribes. It is said that he was difficult to get on with and his writings were abstruse, but were influential in science and philosophy. Bacon's scientific work was very limited, but he was an advocate of the accumulation of data.

Bateson, William (1861–1926). Born in Whitby, and although not regarded as a youth of any potential while at Rugby school, he gained a 1st Class degree at Cambridge in 1883. He then went to the USA to work and returned to the UK where, after some years teaching, he became director of the John Innes Institute. He was married to Caroline Durham and they had three children.

Bauer, Georg (1494–1555). Often called Agricola, he was born in Glauchau, Saxony and became the rector Zwickau school. Although for a while he was a practising physician, his interest in mineralogy led him to become a mining engineer and pioneering mineralogist.

Bede, Venerable (c. 673–735). Bede was born near Weremouth, Durham. At the age of about seven, he was given over to the care of the monastery of Wearmouth and then in 682 he moved to the monastery at Jarrow, becoming an ordained priest in 703. In his new position he became a renowned Anglo-Saxon scholar, theologian and historian. In 1899, he was ordained.

Beringer, Johann Bartholomew Adam (1667–1738). Born in Würzburg, he attended the university there, studying medicine. At the age of 27 he was appointed professor at the university and a year later became the keeper of the university botanical gardens.

Beswick, Thomas (1753–1828) Beswick was born in Northumberland, the son of tenant farmers. He was the eldest of eight children. Although not academically gifted, he was quickly recognised as a skilled artist. When he left school at 14, he was apprenticed to a local engraver. He developed from an engraver of baronial cutlery to book illustrations. He became a partner in the engraving works in 1776, developing the reputation of the company as the finest of engravers. As well as his feelings against war, his moral code towards animals was exemplary, campaigning against docking of horses' tails and cruelty to circus animals. Many of these issues appear in allegorical form in his engravings. He married in 1786 and had four children.

Black, Davidson (1884–1934). Born in Toronto, he graduated in medicine in 1906 from the University of Toronto. He continued there as an instructor in anatomy. Between 1917 and 1919, he was in the Canadian Army as a medic, after which he went to work in Peking (Wade-Giles), becoming head of anatomy in 1924. In 1913, he married Adena Nevitt and they had two children, both being born in Beijing.

Boniface VIII (c. 1235–1303). His non-ecclesiastical name was Benedetto Caetani. He became Cardinal in 1291, being elected Pope in 1294. It was his intention to reassert Papal authority over temporal powers, such as Edward I and Philip IV of France, who disregarded the papal bulls. Boniface VIII was briefly kidnapped in 1303 by the French at Anagni. Shortly after being released he died in Rome.

Bourguet, Louis (1678–1742).   Born in Nimes, Bouruet was a polymath, writing on many different subjects. He was Professor of Philosophy and Mathematics at Neuchatel.

Boyle, Robert (1627–1691).   The youngest of 14 children of the Earl of Cork, Robert was born in Ireland. Good at languages and algebra, he was tutored at home and then went to school at Eton. He travelled through Europe for six years and came back to his family estate in Dorset. In 1654, he moved to Oxford where Robert Hooke was his assistant. He was a director of the East India Company, an alchemist and a founder of the Royal Society.

Brongniart, Adolphe (1801–1876).   The son of geologist Alexandre Brongniart, Adolphe was a French botanist, born in Sèvres, France. He travelled widely in Europe while in his teens, with his father. In 1833, he became professor of botany at the National Museum of Natural History, Paris, a position that he held for the rest of his life.

Bronn, Heinrich Georg (1800–1862).   Bronn was born in what is now Heidelberg, where he attended the university. In 1837, he was appointed Professor of Zoology and became the head of the first Institute of Zoology at Heidelberg. He proposed a tree of life as a method demonstrating graphically the relationships between species and groups. He married Luise Penzel and they had five children.

Broom, Robert (1866–1951).   Born in Paisley, Scotland, Broom studied medicine at Glasgow University. After graduating in 1889, he travelled to Australia and then on to South Africa in 1897 where he settled. From 1903 to 1910, he was professor of Zoology and Geology at Victoria College, losing the position for advocating evolution. In 1934, he gave up medicine and was appointed palaeontologist at Transvaal Museum, Pretoria. He married Mary Baillie in 1893.

Buckland, William (1784–1856).   Buckland was born in Axminster and educated in Tiverton before going to Winchester College after which he won a scholarship to Corpus Christi Oxford. He was ordained as a priest in 1809 and in 1813 he was appointed reader in mineralogy as well. He married in 1825 a girl from Abingdon, Mary Morland, who was an accomplished illustrator in her own right. They had nine children, five of which survived to adulthood.

Buffon, Georges-Louis Leclerc Comte de (1707–1788).   Born in Montbard, Burgundy, Buffon was left a considerable sum at the age of 7 by his uncle. Buffon attended the Jesuit college of Godrans in Dijon from the age of 10 and then, from 1723 he studied law at Dijon. In 1728, Buffon left Dijon to study mathematics and medicine at the University of Angers. With a large fortune he set himself up in Paris to study scientific subjects, initially mathematics but later the biological world. He was married and succeeded by a single child, a son. During the French Revolution, his tomb was ransacked for lead to make bullets.

Burman, John.   Little is known of Burman other than he joined Robert Plot as part of his family when Plot married Rebecca Burman soon after resigning his position at Oxford in 1690. Thus John became his stepson. In the second edition of Plot's *Natural History of Oxfordshire*, Burman made several additions.

Burnet, Thomas (c. 1635–1715). Born in Croft, Yorkshire, Burnet was a clergy man who worked in the court of William III. He had to leave the position when his written work treated the account of The Fall as an allegory, not a literal story.

Butler, Samuel (1835–1902). Born in Nottinghamshire and educated at Shrewsbury and St John's College Cambridge, he gave up the idea of joining the clergy and became a sheep farmer in New Zealand, leaving England in 1859 and returning in 1864. He painted and composed music as well as writing extensively. His financial independence was assured by a legacy.

Chambers, Robert (1802–1871). Born in Peebles, Scotland, he joined with his brother, William, as a bookseller in Edinburgh. He developed his writing in his spare time, becoming a prolific producer of written work. In 1832, they formed the publishing house of W and R Chambers.

Charriere, Henri (1906–1973). Born in the Ardeche, in 1931 he was convicted of murder and was transported to the French penal settlements of South America. After several failed attempts to escape in 1941, he finally made a successful escape to British Guiana (now Republic of Guyana). He eventually settled in Venezuela where he produced his now famous life story *Papillon,* although its veracity has been disputed. He died in Spain of oesophageal cancer.

Collini, Alessandro Cosimo (1727–1806). Born at Manneheim into a wealthy famile, Collini was for many years secretary to Voltaire. He became the director of the Cabinet of Natural History in Manneheim and was the person who described the pterosaur which Cuvier later recognised as a flying reptile. He defended the Manneheim collection against destruction during the French revolution and some years later had it all transferred to Munich.

Colonna, Fabio (1567–1640). A naturalist and botanist, he was proficient in both Greek and Latin before attending the University of Naples where he graduated in law in 1589. Unable to practice law due to epilepsy, he turned his attention to natural history and botany. During the period 1606–1616, he studied fossils. He corresponded with Galileo and he invented a stringed instrument, the *pentecontachordon,* which had 50 strings.

Conybeare, William (1787–1857). Born in London and educated at Westminster School and in 1805 went to Christ Church, Oxford, studying classics and mathematics. Upon graduating he became a curate at various sites, culminating in the position of dean of Llandaff. His interest in geology continued throughout his life and was the subject of his election to FRS. He married once and had several children.

Cope, Edward Drinker (1840–1897). Although well educated, Cope did not excel at school. His father installed him as a farmer, but he preferred a scientific career and attended University of Pennsylvania where he studied biological subjects. During the civil war of North America, Cope travelled in Europe, returning to Philadelphia in 1864, where he married Annie Pim in 1865. They had one daughter, Julia. Being involved with antagonistic competition with Othniel Marsh, Cope gradually lost much of his money, although he was never bankrupt.

Courtney-Latimer, Marjorie (1907–2004). Born in East London, in the Eastern Cape of South Africa, Courtney-Latimer started training as a nurse, but gave up to take a job at East London Museum, South Africa, where she stayed for her entire career.

Crichton, Michael (1942–2008). Born in Chicago and brought up in New York, he attended Harvard and graduated in biological anthropology. During this time he was already writing fiction. He attended Harvard Medical School and graduated in 1969, although he never practised medicine as his expressed intention had always been to become a writer and by this time had been published many times, not always under his own name. He was married many times – Joan Radam (1965–1970), Kathy St Johns (1978–1980), Suzanne Childs (1981–1983), Anne-Marie Martin (1987–2003) and Sherri Alexander who he married in 2005.

Cuvier, Georges (1769–1832). Cuvier was born in Montbeliard, the son of a soldier. He was educated in Stuttgart, originally for the Ministry, but a period as tutor to a family in Normandy confirmed his interest in natural history. Frome 1795 he taught in Paris at the Museum of Natural History. He became Baron Cuvier in 1831 and in 1832 became Minister of the Interior. His work in natural history helped to extend and establish the work of Linnaeus with the addition of the phylum to taxonomy. He was the first to classify fossil mammals and reptiles. He is probably best known for his skill in comparative anatomy, being able to relate organisms by their structure and in being able to reconstruct an animal from only a few key bones.

Daguerre, Louis (1787–1851). Daguerre was born in France and started work as an apprentice to Pierre Prevost, a noted painter of panoramas. Daguerre joined Niepce, who had a basic method of photography, in 1829 and went on to develop the system after Niepce died, turning it into the daguerreotype. In exchange for a lifetime pension Daguerre sold the technique to the French government who then published the method, thereby making it generally available. Daguerre has his name inscribed as 1 of 72, on the base of the Eiffel Tower.

Dart, Raymond Arthur (1893–1988). Born in Brisbane, he was the fifth of nine children. He originally went to the University of Queensland as the first intake at the new institution. He graduated in 1914 and the went on to qualify in medicine at University of Sydney in 1917 he served in the army before moving to the University of Witwatersrand in 1922 where he was professor of anatomy. In 1921 he married Dora Tyree from Massachusetts, divorcing in 1934. In 1936 he married Marjorie Frew, librarian at Witswatersrand University, they had two children.

Darwin, Charles Robert (1809–1882).   Born and educated in Shrewsbury, he is most famous for originating the theory of evolution by natural selection. He studied medicine at Edinburgh University during 1825–1827, but then went to Cambridge in 1828 intending to study for the church. It was here he started his studies in natural history in earnest. He travelled aboard HMS Beagle between 1831 and 1836 gaining many samples and writing extensively on his return on biology and geology. In 1839 he married his cousin Emma Wedgewood, moving to Downe in Kent in 1842. The *Origin of Species by Means of Natural Selection* was published in 1859. The Darwin Building of University College, Gower Street, London, is built on the site of Darwin's London house.

Darwin, Erasmus (1731–1802).   Born near Newark and studying both at Cambridge and Edinburgh, he was a popular and skilled physician, originally in Lichfield. He was married twice and settled in Derby, where he founded a philosophical society. He was grandfather of Charles Darwin via his first wife.

da Vinci, Leonardo (1452–1519).   He was born at Vinci in Tuscany, out of wedlock. His early years were spent with his mother and then with his father. This complicated life resulted in Leonardo having 12 half siblings.   Little detail is known of his early life, even though he is regarded as one of the greatest polymaths to have lived and interest in his life and work has never diminished. His early education was spent as an apprentice to a Florentine artist. His fame broadly lies with his artwork, although he worked in many areas of science and engineering. His scientific studies were empirical and his explanations of phenomena less rigorous than accurate, having little use for mathematics. He was famous when he was alive and his fame has never diminished.

Dawkins, Richard (1941–).   Born in Kenya, Dawkins moved to a family farm in the UK when he was eight years old. He went to Oundle Schooland Balliol College, Oxford, graduating in 1962. Working in California for a number of years, Dawkins returned to Oxford and from 1995 to 2008 was Simonyi Professor for the Public Understanding of Science. He has been married three times, Marian Stamp 1967, divorced 1984. Eve Barham, 1984, divorced after the birth of their daughter. Lalla Ward, 1992, separated 2016.

Dawson, Charles (1864–1916).   Born in Lancashire, he was the eldest of three sons. The family moved to Sussex when he was young. He became lawyer, like his father, with his collecting being a hobby. His premature death was due to septicaemia.

Defoe, Daniel (1660–1731).   Originally Daniel Foe, he was born in London of Presbyterian Dissenter parents and educated locally before being educated at a boarding school in Surrey. He was married for 50 years and had eight children. He worked as a merchant for any years and was also embroiled in politics, narrowly avoiding prosecution after the ill-fated Monmouth rebellion of 1685. In 1703 he was held in a pillory for three days and then taken to Newgate Prison, finally being released. He wrote many satirical and political pamphlets and many tracts which are amongst the very first examples of journalism, although he is now primarily remembered for his full-length books. He is buried in Islington, London.

De La Beche, Henry Thomas (1796–1855).   Born in London, he was brought up by his mother in Lyme Regis on the early death of his father. He was a friend of Mary Anning and an avid collector of fossils. He joined the Geological Society, being President in the season 1848/1849. In 1835 he was appointed director of the Geological Survey of Great Britain. He was elected FRS in 1819 and knighted in 1848. He founded the Geological Museum and the Royal School of Mines.

de Maillet, Benoit (1656–1738).   de Maillet was born into a noble family of Catholics where he received a classical education, although he did not attend university. He travelled widely during his life which allowed him to indulge his interest in geology and natural history. His career was as a diplomat, being French Consul General in Cairo where he studied the Pyramids, then went to Tuscanny and finally in 1715 to the Levant, that ill-defined area of the east Mediterranean as far as Syria. From 1722 he produced manuscripts which would result in his seminal work *Telliamed,* his name spelt backwards. This was published in 1748, after his death.

Democritus (c. 460–370 BCE).   Born in Abdera in Thrace, he was a prolific author on many different subjects, although the only known works that survived were fragments on ethics. He developed the atomistic theory of Leucippus and was a considerable influence on his contemporaries. Karl Marx chose Democritus as the subject for his PhD thesis.

Descartes, Rene (1596–1650).   He was born in a small town near Tours, his home town now taking the name Haye-Descartes. He was educated at a Jesuit College between 1604 and 1614. With enough of an inheritance to make him independent, he spent his life travelling. He was a serving soldier in Holland and Hungary for many years, leaving in 1621 and later settled in Holland for 20 years. He was persuaded to become tutor to Queen Christina of Sweden, but within five months he had contracted a lung complaint and died.

Disraeli, Benjamin (1804–1881).   Born in London and educated privately, he was articled to a solicitor. His first novel was published 1826. In 1837 he became MP for Maidstone. While he was an active member of parliament, he continued his writing. He was Chancellor of the Exchequer before becoming Premier. He was married to the widow of another MP, Wyndham Lewis.

Doyle, Arthur Conan (1859–1930).   Born in Edinburgh, he was educated in Edinburgh and in Germany. He studied medicine at Edinburgh and then started a practice in Southsea. He then went to London as an oculist. Both of these activities were financially unsuccessful. It was this which coaxed him to write. His initial Sherlock Holmes stories were serialised in *Strand Magazine* from 1891 to 1893. He was knighted in 1902. He was married twice having five children, two with his first wife and three with his second.

Drake, Frank (1930–).   Born in Chicago, he went to Cornell University on a Navy Reserve Officer Training Corps scholarship, where he studied astronomy. On graduating he was briefly at sea as an electronics officer. He then went to Harvard to study radio astronomy. His first wife was Elizabeth Bell, with whom he had three sons and with his second wife Amahl, two daughters.

Dubois, Marie Eugene Francois Thomas (1858–1940).   Born and raised im Limbourg, he studied medicine at the University of Amsterdam and taught anatomy at the art schools. To search for human remains he joined the Dutch East Indian Army as a surgeon to get to Java. He married and they had one daughter. In 1897 he was appointed professor of geology at the university.

Durer, Albrecht (1471–1528).   Born in Nuremburg, the son of a Hungarian goldsmith, he was apprenticed to the chief illustrator of the *Nürnberg Chronicle* where he worked until 1490. During the next few years he travelled widely and also married, this presaged a great period of creativity. Although renowned as a great painter, his engravings demonstrate a great sensitivity to the material that he is using and the possibilities and limitations of the technique. He can be seen as the inventor of etching as many of his engravings have lines enhanced by acid corrosion.

Efremov, Ivan.   See **Yefremov**

Empedocles (c. 494–434 BCE).   A native of Sicily, little is known of his life. He was born into a wealthy family and became an extremely accomplished orator. We know that a biography of him was written by Xanthus, but that is all that we know about it because it was lost in antiquity.

Epicurus (341–270 BCE).   Epicurus was born on the Greek island of Samos. His school was well known for eating simple meals and encouraging women members of the school. Although thought to have written a vast amount of material, very little has survived, some letters intact and some fragments. Most of his work is known from later writers reporting on his life. His philosophical ideas all but disappeared throughout the middle ages, finding a revival of interest in the twentieth century.

Eratosthenes (c.276–194 BCE).   Eratosthenes was a Greek mathematician and astronomer born in Cyrene. He was in charge of the library at Alexandria and regarded as the greatest polymath of his time. Besides calculating the circumference of the earth, he wrote on geography and literary criticism. Only fragments of his work remain, secondary sources providing most of the information regarding his accomplishments.

Figuier, Louis (1819–1894).   Born in Montpelier, Figuier trained in medicine and became professor at L'Ecole de Pharmacie of Paris. He wrote extensively on matters scientific and held some extreme views. His wife was a writer of novels.

Fitzroy, Robert (1805–1865).   Born at Ampton Hall in Suffolk. At the age of 12, he entered the Royal Naval College and a year later, he joined the Royal Navy. He commanded the *Beagle* on its five-year voyage with Darwin. Not long after arriving back in 1836, Fitzroy married Mary Henrietta O'Brien, they had four children. After the death of Mary, Fitzroy married Maria Smyth in 1854 and had one daughter. He became Member of Parliament (Tory) for Durham in 1841.

Galilei, Galileo (1564–1642). Born in Pisa, as the son of a musician, he studied music before mathematics and physics. At 25, he was professor of mathematics at Pisa. After he had moved to Padua, Marina Gamba moved in with him. Although they never married, there were three children. He came into direct conflict with the Papacy and at the age of 69 was sentenced to house arrest. His life was filled with mathematical and scientific accomplishments for which he is justifiably remembered.

Gesner, Conrad (1516–1565). Born in poverty in Zürich, Switzerland, he was sponsored to study theology and languages in France and medicine in Basel, later to become city physician in Zürich. His interests were wide and he is often considered a true polymath, writing on scientific matters as well as languages and bibliography.

Gillray, James (1757–1815). Gillray was born in Chelsea, a trooper's son. It was in about 1784 that he became known as an engraver. His work contained many satirical ideas, for which he became justly famous. For the last four years of his life he was judged to be insane.

Goldschmidt, Richard (1878–1958). Gldschmidt was born in Frankfurt and educated in the classical tradition, gaining a place at Heidelberg University, studying anatomy and zoology. He moved to University of Munich to study nematode histology but became interested in genetics. In 1909 he became professor at Munich and studied sex determination. In 1914 he was stranded in Japan with the outbreak of WWI. Travelling to the USA he was interred, returning to Germany in 1918. As the political situation worsened, he realised it was no longer safe for him in Germany, so in 1936 he moved to the USA becoming professor at Berkley, University of California.

Gosse, Philip (1810–1888) Born in Worcester, Gosse went to North America in 1827. It was in Jamaica that he became a professional naturalist with a particular interest in coastal marine species, a subject which he expanded on when he returned to the UK.

Gutenberg, Johannes Gensfleisch (1400–1468). Born in Mainz, he moved to Strasbourg where it is thought he trained and worked as a goldsmith from 1430 to 1444. About 1450 he entered into a partnership to fund a printing press. Five years later the partnership ended and his partner received the printing equipment in lieu of the unpaid debt. Gutenberg is credited with the invention of printing, although the details are not clear, his early products and development of the art were seminal.

Haeckel, Ernst (1834–1919). Haeckel was born in Potsdam, studying in Würzburg, Berlin and Vienna and becoming professor of Zoology at Jena between

1862 and 1909. He travelled widely and wrote on many different biological subjects, being one of the first to produce a genealogical tree of life.

Hawkins, Benjamin Waterhouse (1807–1894).   Born in Bloomsbury, London, Benjamin was the son of an artist. His early education was in art and sculpture, only in his 20s did he become interested in natural history and geology. Hawkins was married to Mary Green in 1826, but left her and their children to enter a bigamous marriage with Francis Keenan with whom he had more children. After the death of Mary in 1880, he re-married Francis Keenan, mainly to legitimise their children. Francis died in 1884.

Herodotus (c.485–425 BCE).   A Greek philosopher, Herodotus was born at Halicarnassus. He travelled extensively in and around Greece and the Mediterranean. He collected considerable information regarding history, geography and social motivations. This information formed the basis for his *Histories*. So comprehensive was his work that Cicero called him 'the father of history'.

Hokusai, Katsushika (1760–1849).   Born in Tokyo, he was apprenticed to a wood-engraver where he learnt the traditional commerative engraving technique, *surimono*. He quickly went his own way, moving into the more modern *Ukiyoye* style. His skill and versatility were renowned and he studied Dutch painting before his most famous illustrations were produced, from 1823 onwards.

Homer.   Homer is the name of an author of unknown provenance and debatable existence. It is not even certain whether the works attributed to the name were even written by the same person.

Hooke, Robert (1635–1703).   Born on the Isle of Wight, he moved to London in 1660 and in 1662 helped found the Royal Society. During the 1660s he formulated what we now know as Hooke's law, dealing with the elastic limit of materials, and realised a spiral spring could control a clock, although it was Huygens who produced the first working model. In 1665 he published *Micrographia* describing the compound microscope and coined the use of the word cell in the biological sense. He was unrivalled as an improver of instruments, such as the microscope, barometer and telescope. Although he was greatly respected, it is said that his cantankerous nature made him difficult to deal with.

Hooker, Joseph Dalton (1817–1911).   Born in Halesworth, Suffolk, Hooker was educated at Glasgow High School and Glasgow University where he studied medicine. His interests were botanical and after many fruitful expeditions overseas took over as director of the Royal Botanic Gardens at Kew, a position his father had held. He married Francis Henslow in 1851 and had seven children, two years after her death in 1874, he married Lady Hyacinth Jardine (this could be an example of nominative determinism) and had two sons.

Hubbard, Bernard (1888–1962).   Born in San Francisco, he studied at several American Jesuit colleges and seminaries before moving to Innsbruck in Austria where he studied theology and was ordained a priest in 1923. Returning to the USA he taught theology and geology, regularly travelling in Alaska. He was often criticised for his inaccurate geology, a similar reception was found when he started anthropological studies in Alaska. He was a renowned and popular lecturer.

Hutton, James (1726–1797).   Born in Edinburgh, he was originally apprenticed to a lawyer, but moved to the continent and studied medicine there, graduating at Leyden. He never practised as a doctor, returning to Scotland to take up farming. This did not hold him and for 14 years he amassed a large sum by extracting ammonium chloride from soot. This financial independence allowed him to return to Edinburgh and devote himself to science.

Huxley, Thomas Henry (1825–1895).   Huxley was born in Ealing, the son of a school master, nonetheless he only received two years of formal instruction, being mainly self-taught. Having studied medicine at Charing Cross Hospital, he entered the Royal Navy Medical Service. After a four-year voyage around Australia he returned to become a self-employed science writer in 1850, and by 1854 was lecturing on natural history at the School of Mines. He worked extensivel in zoology and palaeontology. His employment at the school of mines made him sufficiently financially stable for him to marry his Australian girl friend from eight years previously. Of their seven children there was Sir Julian (biologist) Aldous (writer) and Sir Andrew (Nobel Laureate 1963).

Johnson, Samuel (1709–1784).   Always referred to as Dr. Johnson, he was a lexicographer, writer and critic. Born in Lichfield, Staffordshire, he was a keen reader from an early age in his father's bookshop. He was educated at Lichfield Grammar School and Pembroke College, Oxford. After a period as a teacher, he moved to London where he made a living as a writer. He married a widow 20 years his senior, who died in 1752, plunging Johnson into depression. In 1755 his famous dictionary was published and although a lasting success, Johnson was short of money. In 1760 he was granted £300 a year for life by George III which gave him financial security for the first time.

Kepler, Johannes (1571–1630).   Kepler was the son of a mercenary. He studied theology at Tubingen, but was more interested in mathematics. It was mathematics that he taught at a protestant seminary in Graz. He was forced out of his position in 1600 on religious grounds. As a consequence, he joined Tycho Brahe and being unable to fit observations to the Copernican cosmology formulated his idea of elliptical orbits. He was also a keen astrologer.

Kingsley, Charles (1819–1875).   Born in Holne, Devon, and brought up at Clevelly in the same county. His tertiary education was at King's College, London, and Magdalen College, Cambridge. As an ordained member of the church, he held many ecclesiastical positions, including Chaplin to Queen Victoria. He was also Professor of Modern History at Cambridge University. Two years before he died, he was made canon of Westminster Abbey. He was married and had three children.

Kircher, Athanasius (c. 1602–1680). He was born in Germany, the youngest of nine children, and attended the Jesuit college in nearby Fulda between 1614 and 1618, after which he joined the sect as a novitiate. He studied many subjects besides the standard curriculum, such as vulcanology and Hebrew. He was ordained a priest in 1628 and became Professor of ethics and mathematics at the University of Würzburg, where he also taught Hebrew. He was widely recognised as a polymath, publishing as he did, more than 40 significant works in many different fields.

Kovalevsky, Vladimir (1842–1883). The youngest child of two, Vladimir was born in Belarus. He was educated in languages from a young age and made money translating printed matter while still a student. After graduation he joined the Department of Heraldry, during which he travelled in Europe and eventually settled in London. He married Sofia Korvin-Krukovskaya in 1868 and returned to Russia in 1878, the year that their daughter was born. Unable to gain an academic position and having made some bad business decisions, Kovalevsky committed suicide.

Kuhn, Thomas (1922–1996). Born in Cincinnati in the USA, he graduated with a bachelor's degree in physics, followed by a masters and a PhD, all from Harvard College. After this he moved into history and philosophy off science, becoming professor of the History of Science at the University of California in 1961. Kuhn moved to Princeton University as professor of Philosophy and History of Science and then went to the Massachusetts Institute of Technology as Professor of Philosophy until his retirement. He was married twice and had three children with his first wife.

Lamark, Jean-Baptiste (1744–1829). Lamarck was born in northern France, the 11th child of the family. Upon the death of his father, he joined the French army, at the time stationed in Germany. He received a field commission at the age of 17 with a reduced pension due to injury. He started studying medicine, but gave up tp pursue an interest in botany. In 1778 Lamarck married Marie Delaporte. There were several children, but Marie died in 1792. The following year he married Charlotte Reverdy, who was 30 years younger than him. Charlotte died in 1797. In 1798 Lamarck married Julie Mallet, who died in 1819. In old age Lamarck lost his sight and when he died, the family needed financial assistance for a common grave of five years. After that the grave was dug up and the body disposed of in a lime pit.

Lartet, Edouard Armand Isidore Hippolyte (1801–1871). Born in France, his father was a wealthy landowner. Before taking over the estates, Lartet studied law at Tolouse University. Being of independent means, Lartet spent most of his time in palaeontological investigations. He was made professor of palaeontology at the Jardin des Plantes in the same year he died.

Leakey, Louis Seymour Bazett (1903–1972). The son of East African (now Kenya) missionaries, he studied anthropology at St John's College, Cambridge, and wrote extensively on African anthropology. He married Mary with whom he worked on archaeological digs in Africa.

Leeuwenhoek, Antoinie van (1632–1723).   Born in Delft, he was the first son of five children. He worked as a draper in Amsterdam until 1650 when he moved to Delft. He developd his own techniques for polishing single lenses of very short focal length, which magnified from 50× to 200×.

Leibniz, Gottfried Wilhelm (1646–1716).   Born in Leipzig, Saxony, his father died when Wilhelm was six years old. In 1661, when he was 14 years old, he enrolled at the University of Leipzig. By 1664 he had been awarded a master's degree in philosophy. After one-year study he was awarded a bachelor's degree in law in 1665. In 1666 he gained a doctorate in law (University of Altdorf) and a licence to practice. He was regarded as charming and humorous but due to various personal and diplomatic problems, at his death he was so out of favour with George I and the court that his funeral was poorly attended. His grave in Hanover was unmarked for 50 years after his death.

Leidy, Joseph (1823–1891).   Born in Philadelphia, his father wanted him to become a sign painter, instead he studied medicine at the University of Pennsylvania, graduating in 1844. He went on to become professor of anatomy at University of Pennsylvania and Professor of natural history at Swarthmore College. He married Anna Harden and had one adopted child.

Le Mascrier, Jean-Baptiste (1697–1760).   A French clergyman, he was born in Caen, followed by a traditional theological education. In 1692 he met Benoit de Maillet in Egypt and edited his books.

Leucippus (fifth century BCEE).   A native of Miletus, he was the originator of the atomistic cosmology which was developed by Democritus. It has become almost impossible to separate the ideas of Leucippus from Democritus.

Lhuyd (Lhwyd), Edward (1660–1709).   Born in Shropshire, Lhuyd was the illegitimate son of a comfortably well off family from Wales. He attended and later taught at Oswestry Grammar School. He went to Jesus College, Oxford, in 1682, but did not graduate, joining the Ashmolean Museum as assistant to Robert Plot, later replacing Plot as Keeper. In 1701 he was awarded an honorary MA from Oxford University.

Linnaeus, Carl (1707–1778).   Born in Råshult, Sweden, Carl was the son of a pastor. He studied medicine at Lund, but only briefly and then went on to Uppsala to study botany where he became lecturer in 1730. Having decided to earn a living as a physician, Linnaeus travelled to Holland to qualify in medicine in 1735, the year in which he published *Systema Naturae*, the foundation of modern biological nomenclature. He returned to Sweden in 1738 as a practising physician, becoming professor of medicine and botany at Uppsala University in 1741. In 1749 he introduced the binomial naming system still in use today.

Loew, Friedrich Hermann (1807–1879).   Born in Saxony, he attended the convent school of Rossleben followed by the University of Halle-Wittenberg, where he studied mathematics and natural history. He was appointed lecturer in mathematics and natural history, becoming expert in entomology. He moved to Berlin as a teacher in these subjects. He travelled widely in the near and far east. In 1834 he married the daughter of a notable preacher.

Lyell, Charles (1797–1875).   Lyell was born in Kinnordy, Forfarshire, the son of a mycologist, also named Charles. Educated in Salisbury and Exeter College, Oxford, where he studied law but developed an interest in geology from attending lectures by William Buckland. On graduating he worked as a lawyer. In 1827 Lyell ceased to work in the law and took to full-time geology. During the 1830s he was Professor of Geology at King's College, London.

Lysenko, Trofim Denisovich (1898–1976).   Born in Ukraine, Lysenko initially worked at Kiev Agricultural Institute, developing a completely wrong idea of genetics based loosely on the ideas of Lamarck. The difference between the two was that due to the political turmoil and scientific ignorance of the political leaders of Russia, Lysenko became a leading figure and set back biology in the USSR by half a century

Mantell, Gideon (1790–1852).   Mantell was born in Lewes, Sussex, where he secured an apprenticeship at the age of 15, with a local surgeon. After five years he started his formal medical training in London. Upon graduation he returned to Lewes and started practising as a medic in the local practice. He married Mary Ann Woodhouse in 1816, they had two children, Walter and Hannah. Mary left Gideon in 1839. Having been involved in a carriage accident, Mantell regularly took opium for pain relief and it was an overdose of opium which finally killed him.

Marsh, Othniel Charles (1831–1899).   Born into a farming family in New York state, his education was financially helped by his uncle and a scholarship place at Yale. From there he travelled to Berlin to continue his studies in palaeontology. When he returned to the USA, he was appointed Professor of Vertebrate Palaeontology at Yale, the first such position in the USA. He received a large legacy from his uncle, which made him financially independent and able to pursue his expensive field expeditions. Marsh was a significant protagonist, along with Edward Cope, in the 'bone wars' of the USA.

Mayr, Ernst (1904–2005).   Born in Bavaria, his father died when he was 13 years old, and the family moved to Dresden. He went to university at Greifswald, initially to study medicine but he changed to biology quite quickly. After field trips to New Guinea, he returned to Germany and in 1931 took a position at the American Museum of Natural History. He went to Harvard University in 1953, retiring in 1973 as Emeritus Professor of Zoology. He was married to Margarete, with whom he had two daughters.

Mercati, Michele (1541–1593).   Born in Tuscany, he went to the University of Pisa, studying medicine and philosophy. His important works were not published until more than a century after his death, when his ideas were already being developed.

Mendel, Gregor Johann (1822–1884) A farmer's son, at the age of 21 he entered the Augustinian monastery at Brno, being ordained 4 years later. He studied science for two years at Vienna from 1851, after which he started his experiments in plant hybridisation, which he continued until he became Abbot in 1868. He did not do well at examinations and it was 16 years after his death that the significance of his results was understood.

Meyer, Christian Erich Hermann von (1801–1869). Meyer was born in Frankfurt am Main with club feet, this restricted his movement. He worked initially in a glassworks and then as an apprentice in a bank. Between 1822 and 1827 he studied finance and natural science at Heidelberg, Berlin and Munich. When he returned to Frankfurt, he applied himself to palaeontology. However, in 1837 Meyer joined the Bundestag in the financial administration, which meant that his palaeontology was only carried out in his spare time. To maintain his independence he turned down an offered appointment at Göttingen University.

Miller, Hugh (1802–1856). Born in Cromarty, his father was lost at sea when he was five years old. He was apprenticed to a stone mason at the age of 16, where he worked for 17 years. For the period 1834–1839 he was a bank accountant during which he became disillusioned with the appointments of the Church of Scotland and reinvented himself as a radical journalist, while still writing articles on geology. The cottage in which he was brought up is now a museum run by the Scottish National Trust. Apparently due to illness and overwork Miller committed suicide.

Mullis, Kary (1944–2019). Mullis was born in North Carolina and raised in South Carolina. He studied chemistry at Georgia Institute of Technology and gained a PhD from University of California in biochemistry. He also published in astrophysics while studying for his PhD. He went to several institutions with research fellowships, during one of which he managed a bakery at the same time. Mullis went to the Cetus Corporation where he is credited with inventing the PCR reaction. He was married four times, had three children with two of them, enjoyed surfing and swore a lot, apparently.

Narborough, John (1640–1688). Coming from a Norfolk family, Narborough received a naval commission in 1664, progressing to become Rear Admiral Sir John Narborough. He married once and had two surviving sons.

Newton, Isaac (1642–1727). Born in Lincolnshire, after the death of his father and the re-marriage of his mother, Newton was brought up by his grandmother. He studied at Trinity College, Cambridge, where at the age of 26 he became professor of mathematics. In 1687 he published *Principia*, after which he became more interested in alchemy and theology.

Ostrom, John (1928–2005). Born in New York, he was originally planning to become a physician, but changed his ideas to palaeontology and enrolled at Columbia University. He taught at several colleges until his appointment as Professor at Yale University. He married Nancy Hartman in 1952 and had two daughters.

Othenio Abel (1875–1946).   Abel was born in Vienna, studying law and science at University of Vienna, where he became a professor of palaeontology. His major interest was vertebrate palaeontology, with a neo-Lamarkcian evolutionary slant. During WWII his alignment with the Nazi regime gave him a senior position in Vienna University. This was rescinded after the end of the war when he was forced into retirement.

Owen, Richard (1804–1892).   Born in Lancaster, Owen studied medicine at Edinburgh University and then St Bartholomew's, London. He became the curator of the museum of the Royal College of Surgeons. In 1856 he became the superintendent of the British Museum natural history department. The Natural History Museum was not then an independent entity.

Palaephatus (fourth Century BCE).   This is possibly a pseudonym, and even the original date is uncertain. It is the name given to the author of a set of rationalisations of Greek myths. He put his place as between the believers of Greek myths as literal truths and those who dismissed them as simple stories, although the author does frequently use phrases such as 'this is unbelievable'. The use of the pseudonym may well have been a protection since it would have been seen as heretical to disbeliev the myths in fourth Century BCE Athens.

Paley, William (1743–1805).   Born in Peterborough, Paley became tutor at Christ's College Cambridge in 1768, archdeacon of Carlisle in 1782 and subdean of Lincoln in 1795. He was an advocate for the abolition of slavery.

Peabody George (1795–1869).   Born in Massachusetts, a member of a large and poor family. Initially working in his brother's shop he moved to Baltimore where he developed a business as a financier. He developed a large fortune and used much of it for philanthropic purposes. He died in London and was laid to rest in the USA having been transported across the Atlantic on HMS Monarch.

Pius IX (1792–1878).   Giovanni Maria Mastai Ferretti was born in Senigallia, on the Adriatic coast of Italy. He took Holy Orders in 1818 and by 1827 was Archbishop of Spoleto. In 1840 he was Cardinal and elected Pope in 1846.

Pius XII (1876–1958).   Eugenio Pacelli was born in Rome and was a papal diplomat before being elected Pope in 1939.

Plato (427–347 BCE).   Plato was born in Athens and with the exception of short periods spent most of his life in the city. As far as we know, and somewhat unusually, all of the works of Plato have been preserved, making a considerable collection of philosophical documents. He is generally considered to be the originator of philosophy as it was known in classical terms.

Pliny the Elder (c. 23–79).   A Roman whose name was Gaius Plinius Secundus, which is where the Anglicised name originates. His original training was in law and at the age of about 23 he joined the army as an officer. When he left the army, he lived in Rome where he exercised his legal training. He travelled widely in the Mediterranean basin, becoming familiar with many of the local customs and methods of working. It is reported by Pliny the Younger in a letter to Tacitus, that he died trying to rescue a friend from the eruption of Mount Vesuvius by boat. In the nineteenth century, doubt was thrown on this and some think he died of a heart attack.

Plot, Robert (1640–1696).   Plot was born in Borden in Kent and started his education at Wye Free School, Kent. In 1658 he went to Magdalen Hall in 1658, graduating in 1661. Soon afterwards he developed an interest in natural history. He became the first Professor of Chemistry at Oxford University and the first keeper of the Ashmolean Museum. In 1690 he resigned his position at Oxford University and then married. He became the registrar of the College of Heralds in 1695. He is buried in Borden churchyard.

Ptolemy II Philadelphus (309–222 BCE).   He was responsible for a considerable expansion of the Alexandrian Library and sponsored scientific investigations.

Ray, John (1627–1705).   Born in Essex, his father was a blacksmith and his mother an herbalist. Educated at Cambridge, he stayed on as a teacher after graduation. This was curtailed when he fell fowl of the authorities after the Civil War on the subject of religious observance. He was supported by an ex-student from 1662, touring with him across England and Europe. It was only with Linnaeus that Ray's taxonomy was surpassed. Eventually Ray returned to his home town of Black Notley where he continued writing on a wide range of subjects.

Romer, Alfred Sherwood (1894–1973).   Born in White Plains, New York, he was educated locally, moving to Amherst College where he studied biology. At Columbia he gained an MSc and completed his PhD there. In 1934 he was appointed professor of biology at Harvard University. He married Ruth Hibbard and they had three children.

Russell, Bertrand (1872–1970).   Born in Trelleck, Gwent, he was brought up by his grandmother, educated privately and at Trinity College Cambridge, studying mathematics and philosophy. He was briefly worked as a diplomat in Paris. He married Alys Pearsall Smith, in 1895 and later divorced in 1921, so that he could re-marry. He married his second wife, Dora Black, in 1921 and they were divorced in 1934. His third marriage was to Patricia Spence in 1936 lasted until a divorce in 1952. His fourth wife was Edith Finch, they married in 1952, shortly after his divorce. He was awarded the Nobel Prize for literature in 1950.

Scharf, George Johann (1788–1860).   Born in Bavaria, he went to Munich in 1804 where he trained as an artist. After several years as a miniaturist, Scharf mastered the art of lithography, which was a recent invention. By a strange confluence of events Scharf found himself a member of the English army on the Continent. In 1816 he left to travel to England where he made a living as a lithographer. He worked for Charles Darwin, producing a series of illustrations of fossil material from South America. He was married to Elizabeth Hicks, the sister of his landlady when he originally came to London, and had two sons.

Scheuchzer, Johann Jacob (1672–1733).   Johann was the son of a physician and born in Zürich, where he had his early education. He went to University of Altdorf to study medicine, but finished his medical education at the University of Utrecht. After returning to Zurich he became a town physician and Professor of mathematics in 1710. He was elected FRS in 1704.

Schlotheim, Ernst von (1764–1832).   Born in Ebeleben, Germany, von Schlotheim was initially tutored at home, then attending school in Gotha between 1779 and 1781. He went on to study public administration and natural sciences in Göttingen. In 1792 he entered the civil service in Gotha, achieving Lord High Marshall in 1828. Although his living was as a civil servant, it is for his work in palaeontology for which he is remembered.

Schmerling, Philippe-Charles (1791–1836).   Born in Delft, he studied medicine there and at Leiden after which he was a physician with the Dutch army from 1812 to 1816. In 1821 he married Elizabeth Douglas with whom he had two daughters. In 1822 they moved to Liege where Schmerling continued his studies in medicine.

Schönbein, Christian Friedrich (1799–1868).   Born at Metzingen in Württemberg and at 13 became apprentice in a chemical firm. By private study Schönbein gained a position at the University of Basel, and by 1835 was appointed professor. This was a position he held for the rest of his career. He described the principle of the fuel cell and later discovered ozone. He also described guncotton, nitrocellulose, which it is said he made against his wife's wishes in the kitchen of their home.

Scott, Peter Markham (Sir)(1909–1989).   Born in London, educated at Oundle School and Trinity College Cambridge, where he graduated in history of art in 1931. He went on to study art in Munich and London. He served in the navy during WWII and is credited with the method adopted by the navy for camouflaging ships. He was a founder of the World Wildlife Fund, now called the World Wide Fund for Nature. He was married to novelist Elizabeth Jane Howard in 1942, they had one child and divorced in Divorced in 1951. In 1951 Scott married Philippa Talbot-ponsonby and they had two children.

Senefelder, Aloys (1771–1834).   Born in Prague, he became a successful actor and playwright. His invention of lithography was based upon an observation made around 1796 of drawing on wet surfaces with grease-based material. By 1806 his technique was good enough for him to open a printing works of his own in Munich. He was appointed director of the Royal Printing Office and started a training school as well.

Smith, James Edward (1759–1828).   Born in Norwich, Smith travelled to Edinburgh to study medicine, even though his main interest was botany. When he was 24 the widow of Carl Linnaeus sold him his entire collection of natural history specimens and note books. These were transferred to London where they eventually became the origin of the Linnaean Society.

Spencer, Herbert (1820–1903).   Born in Derby, he had a scientific education from his father but developed an ability to focus on specific subjects to become self-taught in many disciplines. He worked variously as a civil engineer, writer and sub-editor. He moved in the literary and scientific circles of London and started writing works on psychology and philosophy, which were popular and widely read. He is credited with coining the phrase 'survival of the fittest'.

Steno, Nicolaus (1638–1686).   Born in Copenhagen, he was the son of a Goldsmith, due to early illness he grew up in relative isolation. At 19 he enrolled at the University of Copenhagen and after graduating he travelled widely in Europe, settling in Italy in 1666 as Professor of Anatomy at the University of Padua. He was ordained a priest in 1675, later becoming bishop. After his death in Germany in 1686, his body was taken to Italy where he was buried and venerated as a local saint. He formulated four principles of stratigraphy during his lifetime along with much anatomical research. In 1946 his grave was opened, his body removed for a procession through the streets and then reburied.

Strabo (60 BCE–20 CE).   A geographer, he was born in Amasia in Pontus, now mid-northern Turkey, of Greek descent. Most of his life he travelled and studied, settling in Rome after AD14. Although most of his historical works only survive as fragments, his 17-volume work on geography has withstood time and is almost complete.

Strato of Lampsacus (c. 335–269 BCE).   Born in Lampsacus, now in Turkey, Strato was Greek. He attended Aristotle's school in Athens and was a keen student of natural science. Although credited with a considerable body of work, none has survived in the original form and his views are only known through reports by later writers.

Stukeley, William (1687–1765).   Born in Holbeach and educated at Cambridge, he was ordained in 1729, moving to a London ministry in 1747. He carried out extensive field work at Stonehenge and Avebury, but related these monuments to druids.

Talbot, William Henry Fox (1800–1877).   Born at Lacock Abbey in Wiltshire Williamwent to Eton School and Trinity College Cambridge. He wrote on optics and mathematics before moving into photography. Between 1832 and 1835 he was Member of Parliament for Chippenham in Wiltshire. Having excelled at classics as a student, it is little surprise that he was a keen archaeologist and helped decipher cuneiform inscriptions from Mesopotamia.   He was married and had three daughters and a son. He dropped the name William and was usually referred to as Henry Fox Talbot.

Thales (c. 624–545 BCE).   Thales came from Miletus and had a wide-ranging reputation, being a statesman, engineer and astronomer as well as a natural philosopher. He left no writings of his own, other than a star guide, so what is known of his work comes from other sources.

Thomson, Wyville Thomas Charles (1830–1882).   Born in West Lothian, he studied medicine at the University of Edinburgh. Although he graduated, his interests were in natural science. In 1850 he was appointed lecture in botany and a year later, professor at the University of Aberdeen. In 1853 he was apointed professor of natural history at Queen's College, Cork. He changed his name to Charles Wyville Thomson.

Tilesius, Wilhelm Gottlieb (1769–1857). Born in Mühlhausen in northwest Germany. He was introduced to drawing and natural history by his uncle and went on to the University of Leipzig where he studied natural sciences and medicine, graduating in 1795. He was appointed as Professor at Moscow University in 1803. In 1807 he married Olympia von Sitzky, with whom he had a son, but they separated in 1809. In 1814 he returned to Mühlhausen where he continued working although never in a formal position.

Ussher, James (1581–1656). Ussher was born in Dublin, the City where he was educated at Trinity College. In about 1606 he became chancellor of St Patrick's, where he became professor of divinity. By 1602 he was bishop of Meath and in 1625, Archbishop of Armagh. He was a committed royalist, but was favoured by Cromwell. This may be associated with his renowned good temper and charity.

Wallace, Alfred Russel (1823–1913). Born near Usk, South Wales, he was the eighth of nine children. When he was 5, the family moved to Hertford where he went to school until he was 14 and left secondary education. For the next few years he was an apprentice surveyor to his brother William. When William's business declined Alfred left and took up a position of teacher of drawing and map making in Leicester. Following a number of surveying positions Wallace travelled to the Malay Archipelago, collecting and describing species from 1854 to 1862. When he returned to the UK he published articles and popular books on his travels and in 1866 he married Annie Mitten, with whom he had three children.

Warming, Johannes Eugenius Büllow (1841–1924). Usually called Eugene Warming, he was born in Denmark and attended University of Copenhagen studying natural history, where he eventually became professor of botany. He was married to Hanne Jespersen and they had eight children.

Waterston, David (1871–1942). Born in Glasgow, he attended the University of Edinburgh, where he studied for a general degree. He went on to study medicine, graduating in 1895. Upon graduating he became lecturere in anatomy. In 1910 he became Professor of Anatomy at King's College London. In 1914 he went to University of St Andrews, where he was Professor of Anatomy.

White, Gilbert (1720–1793). Born at 'The Wakes' in Selborne, Hampshire, he was educated at Oriel College, Oxford, being ordained in 1747. In 1755 he took the position of curate of Farringdon in Hampshire, while living in Selborne and then latterly curate of Selborne. It was during this period that he wrote *Natural History and Antiquities of Selborne.*

Wilberforce, Samuel (1805–1873). Born in Clapham, he graduated from Oriel College in 1826 and was ordained in 1828. By 1845 he was Dean of Westminster and Bishop of Oxford. In 1869 he was Bishop of Winchester and died after a fall from his horse.

Willoughby, Emily (1986-). Noted for an interest in birds and palaeoart, she studied at Thomas Edison State University.

Woodward, Arthur Smith (1864–1944). Born in Macclesfield, he went to school there and at Owens College in Manchester. In 1882 he joined the Department of Geology at the Natural History Museum where he became Keeper in 1901. He was married to Maud Leanora Ida Seeley.

Woodward, John (c. 1665–1728). Although details of his early life are uncertain, we know he went to London at the age of 16 to be apprenticed to a draper, later studying medicine as an apprentice. It was during this time that he started his collection of fossils. In 1692 he was appointed Gresham Professor of Physic. Upon his death his will gave considerable amounts to Cambridge University for the purchase of land and payment of annual lectures, also his extensive fossil collection.

Worral, Henry (1825–1902). Born in Liverpool, he moved to the USA early in life. Although best known for his artwork, Worral was also a musician, both teaching and composing.

Xanthos of Lydia. (fifth century BCE) Working around the middle of the fifth century BCE, he wrote mainly on history, including a history of Lydia, and occasionally on geology. Only fragments of his original work remain, knowledge of his work comes from later commentators.

Xenarchus of Seleucia (first century BCE) A Greek philosopher who taught at Alexandria, Athens and Rome.

Xenophanes (sixth century BCE). A Greek philosopher and theologian, he was born at Colophon in Ionia, where he lived until he was about 25 after which he went travelling around the Mediterranean, settling for a while in Sicily. His writings in the form of poetry only survive in fragments, which lead us to believe he was an original and independent thinker.

Yefremov, Ivan (1908–1972). Sometimes spelt Efremov, Ivan was born in Vyritsa. Left by his mother in the charge of an aunt who died of typhus, he joined the Red Army. He was discharged in 1921 and went to Petrograd (St Petersburg), where he completed his education. He became interested in palaeontology and became Professor in 1943. It was during the 1940s that he developed the study of taphonomy. He had a parallel career as a writer of fiction with his last work published in 1972. Yefremov was married three times, to Ksenia (divorced), Elena (died) and Taisiya, having one son with his second wife.

Zallinger, Rudolph Franz (1919–1995). Born in Irkutsk, Siberia, and raised in Seattle, he gained a scholarship to Yale University in 1938. After graduating he worked as a painter and teacher. From 1961 until his death he was at the University of Hartford. He was married to Jean Farquharson Day and they had three children.

Introduction

It may seem a slightly circular argument to look at the history of an historical record like palaeontology, but this is what the history of fossils is all about. It is because fossils are so enigmatic, in both origin and wider meaning, that the way they have been interpreted is so important. Such knowledge can tell us a great deal about the social and religious changes that have taken place over the lifetime of humanity. Which, to state the obvious, in geological terms humanity is a split second on the geological time scale of our planet. It is also worth remembering that for many species the fossil record is frustratingly lacking in detail. However, we do know that living animals are a small fraction of all the species that have ever lived. Given such knowledge, it is reasonable to spend considerable time pondering the fossil record, all the progenitors of modern species. It is also reasonable to consider the development of a science that flirts with geology and yet studies ancient species, all of which are extinct.

Before any meaningful discussion could be had regarding the importance of fossils, it had to be agreed what fossils were. It was not always known that they represented organic remains, they are, after all, stone, the most immutable of material. How could a plant or animal be turned to stone without divine intervention? Whether it was to help in developing our perception of ourselves and our position in nature, or to evolution and the origins of life, the apparent anomaly of fossils had to be understood.

It may seem self-evident to us that fossils represent remains of living organisms, but as an untestable hypothesis this is something which requires a great deal of circumstantial evidence to demonstrate as near conclusively as it is possible to get. Part of the historical conundrum which vexed the earliest of observers was that these apparently animal remains, usually bivalve molluscs, were made of stone, and yet no animal has a stone carapace or skeleton. This contradiction between living organisms and fossilised species created a number of explanations of varying plausibility to explain both origin and position. These ranged from high seas

Investigating Fossils: A History of Palaeontology, First Edition. Wilson J. Wall.
© 2021 John Wiley & Sons Ltd. Published 2021 by John Wiley & Sons Ltd.

to account for high-altitude shells, to the rocks being inadvertently seeded with the life force of the found organism that could not develop properly in its rocky environment.

Nowadays we assume that in some way a dead body can become mineralised so that it has the persistence and solidity of stone. It should be understood, though, that fossil material can originate through several different processes of mineralisation, with the final product also varying quite widely, depending on the processes that have taken place. If fossilisation is considered a method of preservation, then most fossils are not well preserved at all, only the shape is retained, a morphological ghost of a living, breeding, physiological organism. Colours are lost with the process of mineralisation as are all the intricate bio-chemical markers that defined the organism. As we shall see, this loss of colour has allowed for wide and sometimes bizarre interpretations, of what a living fossil species might have looked like. This is different to how the animal may have stood or moved. There is adequate information available from fossil skeletons and knowledge of living organisms to give us a clear understanding of the three-dimensional structure of species, but not their colour. There is something else which disappears; behaviour. There is a gap in our knowledge which it is difficult to see ever being filled, and that is the intricate ecology of prehistoric times. Beyond knowing that some species were carnivores while others were herbivores, details which can be inferred from teeth and skeletons, interactions are hidden by time.

While for most the idea of a fossil is of mineralised bones, it does not necessarily have to be so. The meaning of fossil has subtly altered over time from its original Latin root of *fossilis,* which translates as 'obtained by digging' through any found material to its current usage of preserved material, or possibly more accurately, preserved organic material. Using this definition, entrapped organisms in amber also constitute fossils, in this case with much more information preserved than simply a shape; but great caution should be exercised with this idea. The ideas put forward by Michael Crichton in his novel *Jurassic Park* (1990) may be very appealing, but this is just an imaginative fiction. There may well be preserved organic molecules in amber-encapsulated specimens, but within any cell upon death DNA will rapidly start to break up into smaller pieces with increasing entropy. Although there is conflicting evidence regarding DNA decay, preservation in frozen material seems to be best, while long-term preservation in fossilised material seems less and less likely. Work on DNA decay in relatively recently extinct species, such as Moas from the southern hemisphere, has given an insight into the rate of disintegration and associated bacterial contamination (Oskam et al. 2010; Allentoft et al. 2012). All these biochemical analyses and insights are, of course, in addition to the implausibility of a blood-sucking insect of the right type being caught in amber. A better chance of finding longer strands of DNA from ancient extinct species comes with the sometimes ignored natural preservation technique of freezing. This is the situation commonly found in permafrost.

There are some very good cases of preserved remains, most notably mammoths, in permanently frozen ground. The excitement that such discoveries produce is not confined to the past, and neither is it confined to scientific circles. In May 2013, a preserved woolly mammoth was discovered in Siberia and became widely reported, including in the UK press (Daily Mail 2013). So common are mammoth remains found that there is a well-developed trade in the tusks, known as 'ice ivory' and in 2018, there were reported to have been more than 100 tonnes traded, usually for carving into small pieces in the far east. This trade has become so widespread that it has become a concern that poached ivory is being smuggled and traded within the 'ice ivory' trade (Cites 2016, 2019).

The mechanisms of taphonomy, which is the process which results in a mineralised fossil, have been difficult to nail down. This is at least in part because fossilisation is not a single process. A fossil can come about through many different routes, but always involves a set of unlikely conditions and complicated geochemistry. This is well shown by Thomson (2005). We would like to fully understand this process as our knowledge of evolution stands on the foundations of the fossil record. Studying the mechanism of taphonomy is in itself a recent aspect of palaeontology and one which is intimately associated with the geology of the land. It is this linking of conditions and geology which has resulted in such large and highyielding fossil beds as are found in China, Mongolia and the USA.

It is easy to understand that without a process of fossilisation, there would be no fossils and so we could only speculate, devoid of evidence, as to what had lived and breathed on our planet in bygone aeons. In fact, under such circumstances, we may not even consider that there had been species before those that are currently extant. Without the existence of fossils we would have no easily recognised method of aligning the age of the Earth to reality. This would render myths about God creating the planet in seven days and its associated fantasies, difficult to gainsay. However, we do have fossils and so we do recognise the age of the earth as much greater than we can easily comprehend.

It is the human scale of perception of time which made it easy to persuade people that the creation of the Earth happened only a few thousand years ago. By having a version of events which was just about on a conceivable time scale, it made the whole story plausible, which having to think in millions of years was not. Even so, this Biblical explanation involves a period of time which is almost unimaginable when compared to the lifetime of an individual or the social memory of a family. Nonetheless, it is much shorter than thousands of millions of years, which in truth is a scale that really has no meaning to the individual, or even in terms of the life cycle of a species. Species come and go, the average lifespan of a mammalian species seems to be between one and two million years (Mace 1998). This is why palaeontology is so close to geology, it extends so far back in time that only the rocks tell the story, but it is pivotal in the study of the origin and evolution of life.

Part of the modern fascination with fossils stems from the same conundrum with which they were first approached; how can an organic, living being, be converted to rock? The idea that a fossil was the imprint of a living thing made into stone did not gain broad traction within the general population until the nineteenth century, even though it had been voiced prior to this. It is also true that the essential problem of admitting fossils as being mortal remains transmuted into stone was putting science into direct conflict with the Church.

Most early naturalists, and nascent palaeontologists, had gained a classical education, many having been ordained clerics, so there was a long history of reluctance on their part to accept the idea of fossils as organic remains. Such an idea would be in direct conflict with their faith, or at least their education. There was, as a consequence, a long period of published ideas that tried very hard to roll the idea of fossils into a corner where they could be explained away as artefacts or accidental productions of nature, but certainly not the remnants of long extinct species. These arguments were used by Plot (1705) to explain many of the fossils that he found in Oxfordshire.

A change in general attitude towards the nature of fossils took place in the nineteenth century with the parting of the ways between Romanticism and science. Romanticism found its place in the self-indulgent imagination of literature, while the educated imagination of science stood against the idea of assumptions, like those promulgated by religion. It was the broadly untestable ideas of history, which had stood as untested mythical explanations for centuries that were going to yield to the new inquisition of scientific thought. Scientific investigation of fossil material was fuelled by a need to build a consistent picture of the world, which would always be open to challenge as new information arose, but a picture which by its very nature would allow for predictions that could be tested. This was exemplified by the work of Lyell (1832).

Just as we sometimes forget that nobody has seen a living dinosaur, it is also true that nobody has seen the process of fossilisation, that extraordinary series of changes from soft organic to immutable solid. The very indestructibility of stone was part and parcel of the original problem of thinking that fossils could be plant or animal in origin. Even when they were clearly recognised as biological in origin, the questions that were raised were difficult to answer, indeed, they were in some quarters seen as heretical even to ask. The questions immediately associated with fossils were not necessarily at odds with the Biblical story of genesis. For example, if fossils were the remains of animals which were obviously akin to sea animals, how could they be found at the tops of mountains and among cliffs? Several different suggestions were made, the two extremes being that the sea level had dropped (Leonardo Da Vinci, Notebooks 1880) or that earthquakes had raised the mountains (Robert Hooke 1668). Such ideas only held for those species which were obviously marine, problems with skeletons of creatures that ostensibly walked the earth were so mired in difficulty that for the most part such skeletal remains were ignored or put down as unusual rock formations.

Besides the positioning of fossils, there was another problem that vexed both the Church and philosophers. If, as seemed likely, fossils were products of the organic world, both of plant and animal material, they were not like anything currently known. So if they did represent animals or plants, then there would have to be a concept of extinction. This flew in the face of the Bible where creation had taken place as a representation of perfection, and there were neither deletions nor additions of plants or animals possible. As a by-product of this logic, neither were those species that were found, capable of change. If there were extinctions or additions to creation, but most notably extinctions, the implication was that God had made a mistake and had a few attempts at creation before coming up with the final version. Some of this could be partially explained away by assuming that species changed over time. This possibility of a mutable species required a certain flexibility in ecclesiastical interpretation, which was not always possible in some churches.

Even assuming some ability to change, the radical nature of some fossils was difficult to explain in this way. They were just so different, accepting that they had a biological origin would require a big step towards accepting extinction as a real possibility. So another attempt at explaining the animals found fossilised, without treading on ecclesiastical toes, was to suggest that fossils represented species which had died in that locality, but were still extant elsewhere on the planet, or perhaps they had been washed there during the Flood. It was not so difficult to entertain the idea of a species still being alive elsewhere as the planet was ostensibly a much bigger place than it is now. Travel was difficult and slow, indeed there were many areas which remained unexplored until well into the twentieth century. To the explorers of the eighteenth and nineteenth centuries, the immense scale of the Earth left the feeling that there were hiding places where these species, known only from fossils, could be found.

There is one example of this strange phenomenon of a species being known from fossilised remains and then being rediscovered later. Strictly speaking it is a descendent of the assumed extinct lineage which was rediscovered, rather than the same species. This is, of course, the Coelacanth. This astonishing story started in December 1938 when Marjorie Courtney-Latimer was on a trip to the dockside of East London in the Eastern Cape of South Africa, to see what the trawlers had brought up by way of specimens for the East London Museum of which she was a curator. On this particular occasion what she found was a 1.5 m, 57.7 kg fish that was already a day out of water. It had been caught at the mouth of the Chaluma river, just south of East London. Through many trials and tribulations it became apparent that this was a fish representing a lineage that had been supposed extinct for 80 million years. Identification came a few days later as a member of the Coelacanthiformes, a group previously considered extinct. Even more astonishing was that this put it into the Crossopterygii, the otherwise extinct group of lobe-finned fish. This particular species was described and appropriately named *Latimeria chalumnae* (Greenwood 1988). Such a find is a very rare occurrence as it was not a rediscovery of a species, or even just a new discovery of a previously unknown species, this was finding a group of fish that were well known from the fossil record and assumed to have disappeared millions of years previously. If *L. chalumnae* had been discovered in the eighteenth or nineteenth century, this would have been used as a demonstration by biblical literalists and supporters, that the fossil record was of species still extant, but not yet found. There are two points to be made about this, the first is that the oxymoron 'living fossil' is misleading and the other is that as far as we know *L. chalumnae* was not represented in the fossil record, it was the lineage which had survived, rather than the species. Another such surviving ancient lineage is the Tuatara, *Sphenodon punctatus*, from New Zealand. This has been the subject of a quite recent debate regarding the status of a living fossil and the very concept of a 'living fossil' as a useful idea, rather than a confusing one (Vaux et al. 2019).

The situation with the Coelacanth is a very unusual event, although really not so surprising when you consider the numbers involved. As there are generally regarded as being something between 1.5 and 4.5 million species extant on the planet (there are very good biological reasons for this wide range) while through the whole Phanerozoic aeon, which is approximately 541 million years, we have so far found fossils of only about 250 000 different species. This also demonstrates something else; we have no idea of the level of biodiversity for most of the past. We can only surmise what the range of animal and plant species were. There is certainly no reason to believe that the biodiversity was significantly less than it is now. The number and range of species in any ecology will determine the stability as well as the complexity of the population. It is for this reason we have to assume that in the past biological radiations have occurred to fill ecological niches adding to both the complexity and the stability of the system. The complexity and diversity present in the palaeontological ecosystem are reflected in the rate at which dinosaurs are being discovered. In 2019, there were more than 30 new species described and based on mathematical models, this is a small fraction of the potential, with anything up to 70% still to be discovered.

There is no reason to believe that fossilisation is a process only of the past, sedimentation is taking place all the time, with concomitant rock formation, so there should be fossil formation at the same time. Since we do not have a continuous collection of recent fossils from extinct species such as Dodo, Great Auk or Quagga, we can reasonably assume that using a fossil record as a determinant of species numbers is likely to lead us far from the true number. We can, however, make a calculated guess at the diversity from looking at modern ecosystems of the same broad type as we would have expected in any similar area in ancient geological time.

With the developing interest in palaeontology among the general population and the apparently human tendency to collect things, fossils have become ever more important. They have developed an importance, from a scientific point of view, in helping us understand the way species develop and lineages change, or like the clade of Crossopterygian fish, pass through time largely unaltered. Studying fossils can even help with developing completely new ideas and lines of reasoning, such as how birds originated. As part of this they can help immensely in understanding the biggest and most important questions in biology; evolution and the development of modern species.

At the same time, the human urge to collect and make sense of material goods, stamps, coins or fossils has meant that they have become of far greater commercial importance than would have been dreamt of 100 years ago. Wealthy collectors can pay large sums of money for rare and exotic fossils, probably far more than their scientific value merits. This can in itself generate problems, such as scientifically valuable fossils being removed from public view and scientific research. Another problem that can arise is that in the base commercial environment within which we live, it is seen as quite in order to generate chimaeric, or even just plain fraudulent, fossils for sale. Of course, this is not new or confined to fossils. Neither is it a phenomenon limited to areas where the outcome of discovery would be predominantly embarrassment, like faked paintings. When Isaac Newton joined the Royal Mint and supervised the recoinage of Britain in 1696 (there was no paper money), it was estimated that between 10 and 20% of the coins in circulation were counterfeit. Even these examples can be seen as relatively minor compared with the wholesale perversion of scientific knowledge when fraud in science takes place.

Some forms of scientific fraud can be seen as relatively harmless, motivated purely by money, this would be where a fossil has been embellished to make it more valuable, but has little intrinsic palaeontological interest. When wholesale fraud takes place, which disrupts the flow of knowledge, then the fraud becomes far more serious. Strangely, the currency of knowledge is held as of less importance than the currency of gold where this sort of fraud is concerned. Interestingly biological fraud has appeared in many different and sometimes surprising places. An example of financially motivated biological fraud is described in the broadly autobiographical book *Papillon* by Henri Charriere (1970). In this book he describes how, having escaped to British Guiana (now Guyana), he produced an apparently hermaphrodite butterfly by putting male wings on a female body. He further describes how he then sold the specimen to an American for $500.

All this took pace in 1941 when the median annual salary for an American was just under $1000. Even then, during World War II, such productions were a profitable enterprise. It is worth considering the problem of a specimen that is genuine, but is so outlandish that it was thought it might have been a hoax. This was the situation as it occurred with the Platypus, *Ornithorhynchus anatinus*. When a dried specimen was looked at in detail and first described in 1799, it was thought that it might be a hoax. Such composite animals had been known to have been produced by skilful taxidermists in the far east as representations of mythical creatures (Moyal 2004).

In an era when data can be promulgated very quickly, retractions and rebuttals do not necessarily have the same weight as the original message. In these circumstances, fossil fraud can have far greater repercussions than simply questionable science or making dishonest money. Fraudulent fossils may become part of a spurious line of reasoning about creation or evolution and no amount of denial of the obvious lack of veracity of the image of, say, a fossilised giant will counter the belief systems of the ignorant.

Knowledge of fossils cannot be given a start date, as soon as man came into contact with suitable geological areas fossils would have been seen. The cognitive recognition of them being biological in origin may well have come later, certainly as soon as writing became more than simple accountancy. It seems that in the ancient world around the Mediterranean it was easy to slot some of the large bones of fossilised mammals into the mythology, labelling them as being the skeletons of ancient giants and warriors. Because fossils are not confined to one area of the planet, it has taxed the minds of all nations to find an answer to how obviously marine species come to be found so far up mountains. It was only later that people started asking questions about the process of fossilisation itself. The process of fossilisation does take a great deal of understanding and to some extent speculation, since even for short fossilisation periods, it is still far too long when compared to a human lifetime.

Understanding the biological, rather than mineralogical, origin of fossils was a first step in a practical attempt to explain the process of fossilisation. To turn biological material, which everyone knows decays, into a form of rock as solid and stable as any rock, was difficult to comprehend. Once this was understood, at least in part, to be possible, the position of fossils as pointers to the past became undeniable. That we now had a series of items which indicated that species had come and gone, that life had a history and was not brought into being fully formed, made it possible to think of species as being rather more plastic than the ideas of immutability would have had us believe. With that knowledge, species, both fossil and extant, could take their central place in describing evolution.

References

Allentoft, M.E., Collins, M., Harker, D. et al. (2012). The half-life of DNA in bone: measuring decay kinetics in 158 dated fossils. *Proceedings of the Royal Society. B* 279 (1748): 4724–4733.

Charriere, H. (1970). *Papillon*. Rupert Hart-Davis Ltd.

Chrichton, M. (1990). *Jurassic Park*. USA: Alfred A. Knopf.

CITES (2016). 17th Conference Report Johannesburg (South Africa) 24th September – 5th October 2016. *Identification of Elephant and Mammoth Ivory in Trade*.

CITES (2019). 18th Conference Report Colombo (Sri Lanka) 23May – 3rd June 2019. *Consideration of Proposals for Amendment of Appendices I and II*.

Da Vinci, L. (1880). *Notebooks of Leonardo Da Vinci* (ed. J.P. Richter). Volume 2 section XVII Topographical Notes (1880). Reprinted 1888 edition pages 223-270 Dover Publications 1970, New York, USA.

Daily Mail (2013). Preserved Wooly Mammoth from Siberia. *Daily Mail* (9 July 2013).

Greenwood, P.H. (1988). *A Living Fossil Fish. The Coelocanth*. London: British Museum (Natural History).

Hooke, R. (1668). *Discourse on Earthquakes*. St Pauls, London: Richard Waller.

Lyell, C. (1832). *Principles of Geology*. London: John Murray.

Mace, G. (1998). Getting the measure of extinction. *People & the Planet* 7 (4): 9.

Moyal, A. (2004). *Platypus*. Baltimore, USA: The John Hopkins University Press.

Oskam, C.L., Haile, J., McLay, E. et al. (2010). Fossil avian eggshell preserves ancient DNA. *Proceedings of the Royal Society. B* 277 (1690): 1991–2000.

Plot, R. (1705). *The Natural History of Oxford-Shire: Being An Essay Towards the Natural History of England*, 2e. London: C. Brome.

Thomson, K. (2005). *Fossils: A Very Short Introduction*. Oxford University Press.

Vaux, F., Morgan-Richards, M., Daly, E.E., and Trewick, S.A. (2019). Tuatara and a new morphometric dataset for Rhynchocephalia: comments on Herrera-Flores et al. *Palaeontology* 62 (2): 321–334.

1

How are Fossils Formed?

Part of the complex relationship which society has had over the centuries with fossils is at least in part associated with the conceptual problem of exactly how fossils are formed. It was not always assumed that these structures were plant or animal in origin, for a very good reason. From the earliest years of a monotheistic culture, the mortal remains were seen as disposable, epitomised by the *Book of Common Prayer* of 1662 where the funeral oratory includes the well-known 'earth to earth, ashes to ashes, dust to dust' indicating almost by redundant usage that mortal remains will not survive in any shape or form. So it was naturally assumed that with this authority, everything would disappear, and if nothing remained, those stone-like inclusions within rocks could not possibly be animal or plant in origin.

Although inadvertently, the *Book of Common Prayer* reflects something which should be obvious; that fossils are rare. Looking at this from the other direction, it implies that the process of fossilisation is a rare event, and consequently the chances of a specific plant or animal being fossilised are vanishingly small. It took a long time before we understood enough about chemistry that we could have a reasonable idea of how fossilisation takes place.

Fossilisation is a result of a set of conditions which have to be just right to work. It does not necessarily work perfectly every time, and the final product will not always be made of the same material. As we will see later in this chapter, the processes which create fossils vary considerably in detail, which is why fossils also vary so much in their structure and appearance.

The process of fossilisation has to start with the realisation that any living organism is using energy to create a state of order which has to be maintained against the inevitable nature of entropy. Once dead, this process starts to reverse as the organism starts to decay. In some cases, especially vegetable material, decomposition will start with autolysis. It was an understanding of this process in tomatoes that allowed for a genetic modification which considerably increased

Investigating Fossils: A History of Palaeontology, First Edition. Wilson J. Wall.
© 2021 John Wiley & Sons Ltd. Published 2021 by John Wiley & Sons Ltd.

the shelf life by inserting an antisense copy of the 'ripening' gene. The result was the Flav Savr tomato, which appeared for only a few years after 1994. Regardless of autolysis happening, other organisms from large scavengers to bacteria cascade the stored energy of the sun downwards, using it to build themselves up and recycle basic biological materials. In this process, the dead organism reverts to a chaotic state of maximum entropy. Needless to say, this process needs to be halted as soon as possible if any imprint of the dead organism is to be left behind. To cover this process of death and decay through to fossilisation, the word *taphonomy* was coined by I. A. Efremov (1940). He described this in a paper which gave ideas and supplied explanations for the reasons that remains would move from the biosphere to the lithosphere. The meaning has shifted slightly and broadened out in emphasis so that in the twenty-first century, taphonomy covers virtually the entire process of death and decay, with or without any final process of fossilisation.

Before the advent of geochemistry, first described by Christian Schönbein in 1838 (Kragh 2008), and for many years afterwards, there was little by way of a clear idea of changes that can take place in the chemistry of rocks and fossils. It was for many years a simple study of chemical composition of rocks, rather than changes in composition of rocks. This lack of clarity of what might be taking place in the fossilisation process meant that any attempt to describe the process was really a descriptive process of observed events. This was the situation when Charles Lyell (1832) was writing *Principles of Geology*. In grappling with the questions of fossil formation, Lyell expends considerable effort in explaining how various phenomena can result in biological material of all sorts and can become frozen in time. The explanations all stop at the point of 'inhumation', but have an interesting historical context, with descriptions of many examples. These range from inundations by rivers and landslips, such as the draining of a lake in Vermont, USA, in 1810, and the burying of villages when the mountain of Piz in Italy fell in 1772, through to blown sand in Africa. The examples cover many different natural causes of burial, by way of explaining how plant and animal material could move in to the geological strata. At the same time, there is no attempt to describe a mechanism by which this buried material could be changed from biological material, essentially organic, to stone, essentially inorganic, while still retaining some structure of the original organism.

There are exceptions to the normal process of fossilisation, which may not at first even appear to be fossilisation in the popular imagination. These are pickling, freezing, amber and tar pits.

Now commonly used for jewellery, amber is an ancient, preserved, product in its own right. This vegetable product is unique in sometimes containing inclusions of plant material from another species or animal material which can be part or whole small species. Commonly, pollen and plant seeds are found embedded in amber, while the most common animal inclusions are insects, although vertebrates such as lizards have been found trapped in amber. There have been fictional works based on the premise that DNA could be removed from one of these trapped organisms. In such a fictional world, this DNA would then be cloned to produce a new version of the original animal. The most well known of these stories is *Jurassic Park* by Michael Crichton which was published in 1990. This very well thought out story has the quirk of not cloning the animal trapped in amber, but the animal it had fed on. It involved removal of DNA from the gut of an encapsulated mosquito. Supposedly having fed on blood from a dinosaur, the DNA from the mosquito gut was then transferred to a reptile egg which finally hatched as a dinosaur. Sadly, or perhaps not, any DNA in such inclusions would be so badly degraded that it would not be possible to carry the experiment through to the suggested conclusion. Even if all the DNA was present, it would be in short sections, and it is not enough to have the sequence, it has to be in the right order, combined into the correct number of chromosomes. The inclusions within amber, however, have proved very useful in the detailed investigations of arthropod anatomy.

Amber is generally from the Cretaceous period or later, and is mostly composed of mixed tree resins which are soluble in non-polar solvents such as alcohols and ethers. Also present are some resins which are not soluble in the same solvents or are of very low solubility. Most of the resin is made up of long-chain hydrocarbons with groups that are eminently suitable for polymerisation. It is the natural process of polymerisation which causes the change from highly viscous liquid to solid. The process carries on within the solid form and eventually produces a substance that we would recognise as the brittle solid, amber. It should not be considered so unusual that inclusions are found within amber as the amount we know of is really quite large and some of the individual pieces far bigger than we can imagine being produced by modern trees. Precisely why this is so remains a mystery, but so far the largest known piece of amber resides at the Natural History Museum in London and weighs 15.25 kg. To produce such a large volume of resin and then to have it preserved is quite extraordinary. It was originally considered that amber was an amorphous material, which considering its origin and chemistry is a quite reasonable assumption. More recently, it has become apparent through X-ray diffraction studies that in some samples there is a crystalline structure.

The natural process of polymerisation takes place over several years, generally at high temperature and pressure. Just like the formation of all fossils, the process of converting resin into amber is one which is fraught with improbabilities. The original resin has to be resistant to mechanical and biological decay for quite long periods of time, which many plant resins are not, so that there is time for the polymerisation to take place. This will render the resin more resistant to decay or destruction, but does not instantly produce the finished product. These conditions are similar to those thought to be needed for creation of coal, so it is hardly surprising that amber can be found in coal seams.

Initial polymerisation at high temperature and pressure turns the resins into copal. This is a term which originally only applied to resins from South America, but then became a general term for the halfway house between resin and amber. Copal can be used to make a very high quality varnish when mixed with suitable solvents. During the eighteenth and nineteenth centuries, large quantities of copal were consumed specifically to be used as varnish, as it could be applied to any subject that needed a high gloss clear varnish, from carriages to paintings. To complete the polymerisation and to turn the intermediate copal into amber, the pressure and temperature have to be continued. If the pressure is for some reason reduced, but the temperature maintained, the amber, or nascent amber, will break down into its constituent chemicals. In the final stages of polymerisation to make amber, the solvent terpenes are driven off leaving the tree resin as a complex polymer of great resilience.

As one would expect of a product that originates from trees at a time of massive forestation, the distribution of amber is worldwide but heterogeneous in species origin. The majority of amber is generally regarded as being cretaceous or of a more recent in age, which at 142 million years ago, or less, corresponds with the proliferation of flowering plants. Since not all trees produce free resin, it is not so surprising that amber seems to be associated with specific botanical families, of which there are still extant living examples. This is even though the plant families of interest are both ancient and not necessarily flowering. The three family groups that seem to have produced most amber are:

- Araucariaceae, these include the monkey puzzle trees and the kauri trees of New Zealand. They are large evergreen trees which are now almost exclusively found in the wild in the southern hemisphere, but when they were one of the dominant tree species, they were worldwide in distribution. In parts of Turkey, fossilised wood from members of the Araucariaceae is carved and used in jewellery.
- Fabaceae, although most of these legumes are herbs and edible crops, there are some large trees in the family. There is a single tree species in east Africa from which copal is used as incense. They have a widely distributed fossil record, as flowers and pollen as well as leaves.
- Sciadopityaceae, there is only a single species left in this family, the Japanese Umbrella Pine. Although there are no close living relatives, this was a widespread clade with a fossil record extending back more than 200 million years.

It should be emphasised that these are not the trees which originated amber, they are not 'living fossils', they are the current species of the lineage that produced most of the amber we know today. With amber being strictly plant in origin, it should not be a surprise that it is a frequent inclusion in some forms of coal, which were laid down from plant material at more or less the same period as amber was being formed.

Although we all have an idea of the colour of amber, having given its name to the shade of orange which we describe as amber, this is only the commonest of the colours associated with it. For example, there is a form of amber which comes from the Dominican Republic that is quite different. In this form, Dominican amber is predominantly blue. The colour is thought to originate from inclusion in the amber of a molecule called perylene. This is a polycyclic aromatic hydrocarbon with the empirical equation of $C_{20}H_{12}$. Perylene is basically two naphthalene molecules joined by two carbon/carbon bonds. The molecule itself is not blue, but fluoresces shades of blue when illuminated with ultraviolet light, depending upon the wavelength of the ultraviolet radiation. As it is sensitive to a wide range of wavelengths and, of course, ultra violet light is a normal component of daylight, under natural conditions the colour will always appear to be the same. Consequently, the amber will look blue in daylight, but less so, if at all, in artificial light. The unusual inclusion of perylene into Dominican amber implies either a different, possibly unique, species of origin or a considerably modified method of creation.

It is not just by the inclusion of animal material in amber that it is possible to preserve organisms in a near life-like form without the mineralisation normally associated with fossilisation. Along with the inclusion of animal material in amber, there are also conditions in which large-scale remains can be preserved for quite long periods of time. One of these which has yielded some quite startling finds is effectively pickling, in some cases with associated freezing. Although, as we shall see, this latter process can be good enough on its own to render stunning levels of preservation of details after death.

The process of pickling involves an organism rapidly finding its way after death into anoxic conditions, as would be expected in a peat bog where the oxygen has been depleted by large-scale organic decay, usually of plant material. This in itself would cause preservation, although it would depend on long-term stability of anaerobic conditions to preserve organisms intact. In the composite system of preservation, if the remains move to the next step, which is freezing, then the entire animal may be kept in very good condition for as long as the climate permits it. This can been seen very clearly in mammoths removed from permafrost where very little decay has taken place over the millennia of entombment in deep frozen condition. This process of preservation by partial chemical treatment followed by freezing could take place almost anywhere that long-term permafrost can be found.

An extreme example of permafrost preservation was demonstrated at the 47th annual dinner of the Explorers Club held at the Roosevelt Hotel, New York, in 1951. Among the various courses was 250 000-year-old mammoth. This was only a taster as supplies were understandably limited, but we do know that it was provided by the Reverend Bernard Hubbard from an animal found at Wooly Cove on Akutan Island. Akutan is one of the Aleutian Islands in Alaska, where glacial permafrost is widespread. A more recent example of permafrost preservation came with the discovery in 2007 of a frozen mammoth calf. This animal, now called Lyuba, is thought to be the best preserved mammoth mummy ever found. Lyuba is a member of the species *Mammuthus primigenius* and died about 41 800 years ago. It is considered most likely that she suffocated in mud during a river crossing and the lactic acid produced by bacteria in her stomach partially pickled her. So well preserved is she that there is identifiable milk in her stomach. It was not certain at the time of the discovery that the calf would be preserved at all, as while the original discoverers went to report the find to their local museum, the corpse was lifted and removed for display outside a shop. It was retrieved, with a little damage due to local dogs, and has been widely exhibited since. The permanent display is at the Shimanovsky Museum and Exhibition Centre in Salekhard, Russia. Salekhard is reputedly the only town where the Arctic Circle actually runs through the town.

Completely submerging an organism in what is in effect a preservative is another way in which plant remains or animal corpses can survive for very long periods. One of the ways this can happen is found in tar pits. These are rare sites, but can be quite extensive, for example, Pitch Lake in Trinidad is the largest natural deposit of asphalt in the world, covering about 44 ha. Other such deposits include Binagadi asphalt lake in urban Baka, Azerbaijan, and the second largest lake, Lake Guanoco in Venezuela.

Probably the most investigated and well known of all the asphalt deposits are the Tar Pits of La Brea in Los Angeles. This is the most well studied of the rare tar pits, and it is also unique in having been an asphalt mine during the nineteenth and early-twentieth century. The position of the tar pits, in what is now an extensive suburban area of Los Angeles, has added to the development of interest in this particular site. The current pits are mostly man-made, a leftover from asphalt mining during 1913 and 1915, and partially created by deliberate digging for bones. It had been realised for a long time prior to the twentieth century that there were bones in considerable numbers present in the tar pits, but they had been assumed to be those of stray cattle that had wandered in to the tar and become stuck. It is true that animals wandered onto the unsupportive and sticky surface, making the mistake of thinking it was a watering hole, what had not been realised was that these wandering animals were not generally domesticated cattle. It would have been an easy mistake to make for wildlife to imagine the surface was solid as it would accumulate twigs and leaves and form pools on the surface when it rains. This would also attract night flying aquatic insects, such as water beetles that would similarly become trapped in the sticky tar. This is a well-known phenomenon, where newly made roads with an apparently wet surface will attract night flying aquatic insects when there is a bright moon.

Once in the tar, escape is extremely difficult for the trapped animal, even if they were to escape, the mammalian behaviour of licking fur would render the animal sick due to the toxic compounds in the tar. It is similarly thought that the large numbers of carnivores which are present are due to them being attracted by the plight of the struggling trapped animal. The two predominating carnivorous species that were trapped in La Brea tar were the Sabre Toothed Cat, *Smilodon fatalis* (Figure 1.1), which is the second commonest skeletal remains of any sort recovered from the tar pits, and the Dire Wolf, *Canis dirus*. There are more than 400 skulls of *Canis dirus* on display at La Brea which have been recovered from the tar.

The lack of preserved soft parts in the tar pits has allowed speculation regarding the coat colour of species of *Smilodon*. They have been represented as plain-coated or spotted, either of which would be possible. The coat colour of modern felids seems to be broadly dependent on the preferred terrain in which they live, but since there are exceptions to this, it becomes impossible to be sure of the coat in these species.

The formation of these tar deposits starts with a natural seepage of oil from underground reservoirs, as it reaches the surface, the lighter fractions evaporate or are used as an energy supply by some of the microorganisms present, leaving the heavy tar behind. Long after it was known that tar pits contained animal remains, it was not understood why only the skeletons remain. Part of the answer is quite prosaic, it seems that it takes a long time for the corpse to sink, quite long enough for decay to take a considerable hold on the soft parts. Besides this,

Figure 1.1 *Smilodon fatalis (californicus)* skull from La Brea Asphalt, Upper Pleistocene Rancho La Brea tar pits, Los Angeles, California, USA. Staining due to the tar renders the bone permanently discoloured. *Source:* Photo. James St John, Creative Commons, generic.

the tar has residual solvents in it which disrupt the lipids in the body. Lipids are a group of organic molecules less related to each other by their chemical structure as by them being soluble in non-polar solvents such as benzene and ether. The other major cellular components, proteins, are not soluble in non-polar solvents, and it is these that would be decayed by fungi and bacteria or scavenged by small insect such as flies.

Although preservation of ancient material, plant or animal, by encapsulation can result in very high resolution remains, the most usual way of thinking about preserved remains is as inclusions within rock. This process requires considerable changes in chemical structure and composition, a process described as taphonomy. The final outcome of taphonomy in the most frequently considered situations of fossilisation may appear to be the same, that is leaving a permanent record set in stone, but it takes little time investigating various fossils to see that the mineral nature of fossils can be radically different. This is most notably so when comparing fossils from different areas, as the colours vary quite widely. These colour variations reflect different mineral compositions within the final product of fossilisation, which of course is a reflection of the mineral composition of the rock in which the fossil was formed.

In broad terms and very simple terms, fossilisation resulting in a stone product requires rapid sedimentation of material which will eventually bind in a cement-like fashion to become rock. The details can, of course, vary enormously from site to site, but in broad terms it always starts with sedimentation. This is one of the reasons that it is generally considered that fossilisation only takes place in shallow seas, lakes or shallow slow rivers and very wet swamp land.

Slow rivers and swamp land are often associated with floodplains, which also accumulate remains washed down stream and silt to cover them. The converse conditions are not so conducive, that is, fossilisation would not normally take place in dry, arid, conditions. This inevitably has some implication for the types of fossils which are most frequently found. Aquatic species will naturally form the bulk of fossilised material, but all species need water to drink and watering holes that attract grazing livestock also attract carnivores, both to drink and as an easy way to gain access to prey species.

A large part of the reason that fossils are not 'dry found' is that although mummification through desiccation is an excellent way of preserving mortal remains, we have to consider the length of time they can survive. The longevity of a mineralised fossil plays very well when considered against survival of mummified remains that are simply desiccated. In dry conditions, scavengers and recycling organisms will be active in breaking down bodies that contain nutrients and valuable mineral resources. Even when desiccated, there are a number of invertebrates that can use the material as a source of food. Assuming that burial takes place and the situation is one in which decay and scavenging do not occur, there is still a major long-term problem of physical stability. In dry conditions, there will generally only be loose compaction of the overlying material, and this implies that there is insufficient mechanical stability of the substrate to guarantee survival of mummified remains. Shifting substrates can be a problem with standard models of fossilisation, unless the fossil is rapidly compacted and incorporated as part of the rock. By contrast, dry mummified remains in loose material, although dry and preserved, will be shifted about and abraded very quickly to dust by the surrounding material. This is very much associated with the time scales which have to be considered when thinking about the age of fossils and the aeons over which they have survived. Taking time to try and comprehend these immense time scales when compared to the age of mankind and associated civilisations is worth the effort, even though it is extremely difficult to appreciate the length of time involved.

It has been suggested that by comparing modern ecosystems with the fossil record, it may be possible to determine the biodiversity and species numbers in extinct ecosystems. By making a range of assumptions, based on ecosystem complexity, it is also possible to estimate the rate at which organisms leave visible traces. Modern studies would indicate that comparable ecosystems, such as rainforests in South America and West Africa, have the same broad biomass divided up into the same numbers of species and individual organisms. By implication, it would seem reasonable that comparable extinct ecosystems would have comparable numbers of species and comparable sizes of populations to modern ecosystems. Needless to say, the species would be radically different, but there would still be primary producers and an energy pyramid leading to the apex predators. Using these broad assumptions we can estimate that, depending upon conditions, anything from 0 to 70% of an ecology can become fossilised, with an average of 30% of biota leaving a trace of some kind. It is a very wide range, which is a reflection of the uncertainty of these sorts of estimates. From these numbers it should be self-evident that most organisms don't leave any trace at all. If there was little or no recycling of organic remains, in our modern forests and woodlands we would, for example, be wading through the annually discarded antlers of deer.

One of the reasons for this low fossilisation rate, besides the unsuitable terrain for the process to take place, is the recycling of biological material. For the soft parts, the organs and muscles, recycling primarily takes the form of being food for other animals, followed by bacterial and fungal decomposition. With skeletons it is a little different, it is the mineral content which is of value to other animals, rather than the calorific food value. We may assume it is the calcium that is the prime object of recycling, but is not the calcium that is the primarily used part of a recycled skeleton, it is the phosphate and some of the organic material.

For many years, it was assumed that there could be little experimental work possible to investigate the formation of fossils. This changed as interest in fossil fuel formation developed with the increasing demand that became apparent throughout the twentieth century. It has been possible to demonstrate that compression in fine sediments, or in the form of fine clay (Saitta et al. 2019) followed by heating, can produce a result that is very similar to fossilisation. The fine sediment leaves enough space for labile hydrocarbon molecules to escape, which has implications for the fossil fuel industry. At the ultrastructural level, artificial encapsulation of organic material has the same appearance as high-quality fossils. Long-term survival of organic residues through aeons of time in the geological record depends primarily on their chemical structure. Hydrophobic water-insoluble organic molecules can last very well (Bills 1926), while some molecules such as proteins and DNA have a relatively short survival time. Proteins and DNA probably have a short survival time in water, or damp conditions, due to thermodynamically unstable phosphodiester and peptide bonds as well as the instability of some amino acids. Deeply embedded short sequences of nucleic acid, as might be found in the teeth, can be extracted from some preserved material. If the material has undergone high temperatures, or prolonged heating, nucleic acids will break up and will only be present as very small fragments and residues.

Although the general perception of a fossil is of an image, almost an engraved image on stone, or sometimes a three-dimensional construction, the process of creating a fossil is not a uniform one. This is perhaps self-evident, since it can hardly be expected that preservation of a heavily calcified mollusc shell would follow the same process as a vertebrate skeleton. There are some parts of the process which are well documented and occur quite commonly. Even the common routes quickly diverge down different paths so that the result is a wide range of fossils being preserved in a variety of different ways.

By far the commonest fossilised forms to be found are those animals that start with a mineralised structure and have high population numbers (Donnovan 1991). These are most obviously molluscs, corals and echinoderms. As a structural element, the most frequently encountered element forming hard parts in the animal kingdom is calcium (Ca). This appears in many forms but is most commonly found among invertebrates in a simple chemical structure, either calcite or aragonite. These are of the same chemical composition, but different crystal structures. In vertebrate skeletons, calcium is conjugated in a different way and forms part of modified hydroxyapatite. This makes up more than 50% of the bone. Hydroxyapatite is a slightly more complicated molecule having the empirical formula which is normally written as $Ca_{10}(PO_4)_6 \cdot 2OH$. As the primary structural calcium salt of vertebrates, both in skeletons and in the teeth, it is this molecule which can have the hydroxyl group replaced by fluorine in the teeth, giving it greater resistance to decay.

In whichever form the calcium is found, it is these calcium deposits which form the basis of the most commonly found fossils, coming as they do from shells and skeletons. This is not to say that they remain unaltered, but only that these hard minerals are the starting point for what can be quite complicated chemical changes that are found in diagenesis.

Both calcite and aragonite have the same empirical formula of $CaCO_3$, but they are differentiated by virtue of the crystal structure that they take up. On close, very close, X-ray diffraction examination, calcite is a hexagonal (trigonal) system and aragonite is rhombic. Structurally the difference is quite small, but aragonite tends to be less physically stable than calcite. It is probably for this reason that calcite is the most commonly found crystalline calcium salt and is therefore the major component of the large deposits of marble and limestone which are so easily seen on some exposed cliffs. When it is found with magnesium carbonate, $MgCO_3$, the mixed marble-like material is described as dolomite.

The distribution of these two forms of calcium carbonate, aragonite and calcite, in biological systems varies depending on the taxa. Although not necessarily exclusive, we would generally expect to find calcite in such wide-ranging groups as brachiopods, ostracods, foraminifera and some sponges. The alternative structure, aragonite, is more often present as the structural material in molluscs and some sponges. That both forms are found in sponges is indicative of there often being a mixed use of calcium as a structural element in the same taxonomic group.

The process of forming what we would most readily recognise as a mineralised fossil involves the process known as diagenesis. In broad terms, diagenesis describes the changes that sedimentary deposits undergo, both chemically and physically as well as changes due to biological activity, before the process of lithification takes over to form solid rock. The first stage of this is permineralisation, which involves the percolation and deposition of crystalline material from solution into areas where only water-based solutes can reach. Because it is a crystallisation process, the fine internal detail can be very well preserved. The level of external detail which is preserved also depends on the type of material in which death has taken place. As would be expected, the finer the sediment, the greater the final detail.

Calcium permineralisation is a common early occurrence in fossil formation, as calcium salts quickly saturate ground water being of relatively low solubility. Aragonite is not often preserved unaltered in the geological record and where aragonite fossils do occur they are usually associated with mudstones and marls. Under suitable conditions, it is possible to find ammonites that still have their aragonite shells, the same is also true for bivalve molluscs, but only if conditions are right for preservation without structural modification. The structure of aragonite is less stable than calcite, so where aragonite is present in shell and skeletal structures, it tends to be replaced by calcite. Generally rhombic aragonite is often found to have been replaced by hexagonal calcite, and this usually takes place in one of two ways. It should be noted that although both of these molecules have the same formula, it would seem that when aragonite is replaced by calcite, the calcium carbonate of the structure is not necessarily reused, the calcite deposits coming from extraneous calcium carbonate dissolved in the surrounding water.

The first process for substituting aragonite for calcite involves complete dissolution of the aragonite followed by deposition of calcite in the void that was left. There can be a considerable time lag between dissolution and complete deposition, this is a purely chemical process, unlike the original formation which was under biological control. If calcite deposition is a slow process, it can result in large crystals of calcite being laid down with the loss of a great deal of the original detail. If there is a gap between the two processes, it is reasonable to assume that this is why the original calcium carbonate is not wholly being reused, if at all. During the time between dissolution of the aragonite and deposition of calcite, the shape of the organism is held in place by a cement layer in the host sediment. This can be either as a complete block, in which case infiltration by depositing solutions will tend to take longer, or as coating laid down on the remains of the organism and fused into a cement case within a looser sediment, under these conditions calcite deposition is faster.

The second method by which aragonite can be replaced by calcite takes place across a thin film, with the aragonite being dissolved on one side of the film and calcite being deposited on the other. This process is often referred to as calcitisation and is a process which can retain considerable skeletal detail in the final fossil. It also tends to reuse more of the original calcium carbonate, although this is not necessary if ambient conditions are correct, for example, if the water is a saturated calcium carbonate solution. With the replacement of one form of calcium carbonate with another, there is an almost inevitable reduction in fidelity since the original structure would have had other biomolecules present, such as scaffold proteins and the crystal structure would be very small indeed, controlled by the cells of the living organism. So laying down a calcite copy by simple chemical crystallisation will follow the original mould but will have larger crystals.

Even though we refer to calcite as a uniform substance, the nature of calcite is such that it is described as having two forms itself. These are Low Magnesium Calcite (LMC) and High Magnesium Calcite (HMC). The difference between these two is quite small. LMC has between 1 and 4 mol% $MgCO_3$, while HMC contains 11–19 mol% $MgCO_3$. This small difference between the two is enough to make HMC less stable than LMC. Consequently, it is possible for skeletal material that was originally laid down by the organism as LMC to be retained through geological time and consequently retaining structural details down to the micro level. This situation is found in the brachiopods, which start with an LMC structural composition and consequently are sometimes little changed during fossilisation. With the original material remaining in situ it can be used to give a good basis for analysis of carbon and oxygen isotope values from their LMC calcite shells. Similarly, trilobites are also well represented in the fossil record, not just because they were a common part of the ecology, but also because their shells were mostly calcite with some phosphates present.

Structural crystalline minerals other than calcium carbonate can occasionally be found to have replaced calcite in some fossils, although this generally only happens in very specific circumstances. It is not always known in taphonomy what these specific circumstances are, but the fact that the substitutions do not take place commonly would imply specific chemical needs for the process to go to completion. It may also require unusual pressure and temperature to complete the conversion. Silification, that is, substituting silicate minerals in place of the calcium based calcite, is something which happens occasionally. In certain circumstances, silification may have spectacular results, one such is when fossils become converted to opal.

Hydrated amorphous silica (SiO_2 nH_2O) in the form of opal can form fossils which are so decorative that they are often found being broken up for use in jewellery. If the silicates enter a body cavity and precipitate from solution, the external structure can be very well preserved, but not the internal details. As an alternative, should it infiltrate the organic material before decomposition, then good internal details can be preserved. Because deposition of silicon depends on the ground water and basal rocks, it is a very rare combination of events which results in opalised fossils. Some of the very best sites for these opal fossils are found in Australia, usually in active opal mining areas, such as Lightning Ridge in New South Wales and Coober Pedy in South Australia. Although there are other sites in the world where opalised fossils can be found, Australia is the most renowned area. It was in 1987 at Coober Pedy in South Australia that what has become known as Eric the Pliosaur was discovered. This reptile, *Umoonasaurus demoscyllus,* is one of the most complete opalised vertebrate skeletons known. Not only is it of interest and value as a fossil, it is also of considerable financial value for its opal content alone. The history of the fossil after being discovered is quite convoluted. It was originally sold to a dealer who went bankrupt, at which point it was very nearly sold to a jewellery consortium which had planned to turn it into jewellery at which point it would have been opal with the enhanced value of being a fossil. With help from Akubra Hats and money raised by school children, the fossil was bought for the Australian Museum, where it is now on show.

The formation of opal fossils is a rare event, just as the formation of opals themselves is. Although opalised fossils must have a quite complicated process of formation, we can gain some insight from the way that gem-stone opals are formed. Opal formation starts as silicon oxide spheres in a silica-rich solution. These spheres settle under gravity and build up to form the gemstone opal. This process is very slow and to produce the precious colours of opal, the spheres need to be uniform in shape and between 150 and 400 nm in diameter. Although opal is generally made from even-shaped and sized spheres, there is no long-range or short-range order in the stacking of the spheres. There are broadly two forms of opal, opal-A (sometimes AG or AN) and opal-C (sometimes CT). Opal-A can transform into opal-C under high pressure from overlying sedimentary deposits. This form of opal can sometimes contain as much as 10% by weight of H_2O.

Another mineralisation process which replaces calcium in fossilisation is pyritisation in which iron salts are laid down instead of calcite. This takes place when the water in which the organism died is high in iron sulphides. When surrounding water is high in iron, deposition of FeS_2 is affected by sulphides originating from decaying organic matter, of which there will be considerable amounts in a corpse. Pyritic fossils can be difficult to store because under humid or damp conditions, the iron pyrites of which they are made up, that is FeS_2, will chemically decay into iron oxide and sulphides, sometimes sulphur itself. This is a process which can be hastened by some types of bacteria. If this does start to happen to a pyritic fossil, it is possible to find sulphur deposits on the surface and the fossil may expand as iron oxide is a much larger molecule. This is exactly the same process which causes rust on ferrous metal to flake as the iron oxide pushes itself apart. Iron pyrite is not the only iron containing mineral which can create a fossil impression, but it is by far the commonest. Pyritisation has been shown to be a process capable of recording considerable morphological detail of soft-bodied organisms of the Ediacaran (Smith 2019).

It is possible for rapid precipitation to occur around a subject being fossilised which will then form a nodule retaining considerable detail in the organism at the centre of the nodule. A mineral that is often found involved in this process of nodule formation is siderite, that is iron (II) carbonate, $FeCO_3$. The unusual thing about siderite is that it is approximately 48% iron and consequently a valuable iron ore for commercial production of steel. Siderite is a diagenetic mineral in shales where it creates authigenic moulds of fossils as nodules, which have to be split to discover their content. One of the most famous sources of these fossil nodules is the Mazon Creek fossil beds in Illinois, USA, where it is even possible to find fossil sharks. These are rare fossils because although shark teeth are common, their skeletons are cartilaginous and therefore do not normally fossilise before decaying.

Interestingly there is a single species of living gastropod, which is the only animal known to use an iron-based mineral for its shell. *Chrysomallon squamiferum* is a gastropod from deep sea thermal vents at depths of 2500 m and greater, where the outer shell contains iron sulphides including the mineral greigite, Fe_3S_4.

While there is a general procedure of one mineral replacing another, or sediment developing in a systematic way over the top of a form which will become fossilised, this is, perhaps, an oversimplification of a dynamic process. It has been shown that the linear progression model may not be so common as a simultaneous process, depending upon tissue type. It is quite likely that local chemistry within the cadaver will alter ion concentrations and influence specific precipitation events. These chemical changes will also be influenced by the state of decay, and it has been suggested that rapid biodegradation can enhance the detail which is eventually left behind (Jauvion et al. 2020). The rapid deposition of some preservative minerals will then be replaced at a slower pace by more robust minerals such as calcite.

As can be appreciated, the formation of fossilised material is a rare chance event which unless interpreted carefully gives a distorted snapshot view of the past. The image which can emerge is one where ecosystems were dominated by the hard shelled or skeletoned. This may broadly be correct, but with the difficulty that soft parts do not generally leave a trace before decay, we may never know the extent or range of some of the species and whole taxons which were on the planet before vertebrates appeared.

References

Bills, C.E. (1926). Fat solvents. *Journal of Biological Chemistry* 67: 279–285.

Chrichton, M. (1990). *Jurassic Park*. USA: Alfred A. Knopf.

Donnovan, S.K. (1991). *The Process of Fossilisation*. Columbia University Press.

Efremov, I.A. (1940). Taphonomy: a new branch of palaeontology. *American Geology* 74: 81–93.

Jauvion, C., Bernard, S., Gueriau, P. et al. (2020). Exceptional preservation requires fast biodegradation: thylacocephalan specimens from Voulte-sur-Rhone (Callovian, Jurassic, France). *Palaeontology* 63 (3): 395–413.

Kragh, H. (2008). From geochemistry to cosmochemistry: the origin of a scientific discipline, 1915–1955. In: *Chemical Sciences in the 20th Century: Bridging Boundaries* (ed. C. Reinhardt). Wiley.

Lyell, C. (1832). *Principles of Geology*, vol. II. London: John Murray.

Saitta, E., Kaye, T., and Vinther, J. (2019). Sediment-encased maturation: a novel method for simulating diagenesis in organic fossil preservation. *Palaeontology* 62 (1): 135–150.

Smith, E.F. (2019). Pyritization of soft tissue at the Precambrian-Cambrian boundary in the southwest USA. *The Palaeontological Association Newsletter* 102: 65–67.

2

Descriptions and Uses of Fossils

To ancient philosopher scientists living in a world of apparently fixed age, which was assumed to be measurable in thousands of years, explaining fossils became something of a problem. This was primarily because fossilised bones and shells, especially shells, are found where they would not be expected. This is even more so when most of the fossilised remains that are being looked at are sea shells at the top of a mountain. It consequently takes considerable imagination to produce an explanation which is both plausible and acceptable for seas shells on hill tops miles from the coast and skeletal remains in solid rock. The easiest way of dealing with anomalies of this sort would be to simply ignore the problem and accept the existence of high-altitude shell beds as a natural phenomenon. But enquiring minds will always look for a unifying hypothesis that can in some way be tested, so that it can be turned into a useable explanation.

In this respect, we should not judge ancient ideas using modern standards or levels of knowledge. While we now know that about 71% of the earth's surface is covered by seas and oceans, this is relatively recent knowledge. When questions were being asked about fossils for the first time with an attempt to produce an explanation of their existence, not just an observation of them, the scale of the earth was too large to be easily conceived. It was also complicated for the Greeks and Romans, since although they knew the boundaries of the Mediterranean, the waters of the Atlantic Ocean in a westerly direction apparently had no end.

The complicated and volatile geology of the Mediterranean also made it quite easy for people to imagine that, however, solid the land upon which they trod appeared to be, it was not inconceivable that it could have been inundated and covered with water. This is opposite to the modern idea of land being thrust out of water by volcanic or tectonic activity. In ancient days, the sea was regarded as capable of rising up and simply covering everything, indeed this was a story which has appeared in many cultures. Instability of the land would have been

Investigating Fossils: A History of Palaeontology, First Edition. Wilson J. Wall.
© 2021 John Wiley & Sons Ltd. Published 2021 by John Wiley & Sons Ltd.

considered normal as the Mediterranean basin has always been volcanically active with the African tectonic plate moving northwards against the Eurasian plate.

In terms of palaeontology as we recognise it now, it is often said that the publication of *De Rerum fossilium, Lapidium et Gemmarum maxime, figuris et similitudinibus Liber* in 1565 by Conrad Gesner (Figure 2.1) marked the beginning of modern palaeontology. This is, however, only of the modern era; puzzling over rocks of unusual shape and composition had been going on for a long time before Gesner wrote his book.

Long before Gesner, and amongst the earliest known speculators regarding fossils, was Xenophanes of Colophon (c. 570–475 BCE). It is important to realise that Xenophanes would almost certainly not have been the first speculator. Such thoughts were probably commonplace wherever fossil beds were revealed, a reflection of natural human curiosity. Earlier speculation undoubtedly took place but was rarely recorded, so it is only the written or reported speculations which we can take as evidence of curiosity regarding the origins of fossils. Xenophanes was a Greek theologian and thinker, living around 570–480 BCE. He is thought to have

Figure 2.1 Frontispiece of *De Rerum fossilium* by Conrad Gesner (1565). The text was written in Latin, as was usual at the time to ensure a wide readership among the educated.

been peripatetic for most of his life, having left his home in Ionia when he was 25 years old. Because there are no complete works by Xenophanes, only remnant fragments, we cannot be sure of the line of logical debate which he took, nor the weight that he put on his own assumptions (Kirahan 2010). What we do know is that he examined fossils, in the form of relatively high-altitude sea shells. It was this that led him to the conclusion that water had once covered the entire planet. To Xenophanes, it would seem to be the only reasonable explanation for why mountains and hills contained marine molluscs. The idea of a mobile earth on the scale necessary would have been ridiculous compared with a notion that the very fluid ocean could have just filled up and inundated the earth. Once the seas receded, islands and the lands around the Mediterranean would have been revealed as dry land. One thing which we do know about Xenophanes is that he was a disciple of Anaximander who had suggested not only that the world arose from a formless material, but that it had once been covered entirely by water (Urmson 1960). Xenophanes developed the ideas of Anaximander further, citing the fossil shells found at high altitude as a clear demonstration of a watery planet in the ancient past.

After Xenophanes, by only a few years, the historian Herodotus (c. 484–425 BCE) also found shells far from the sea and pondered upon the possibilities of how they got there and what their significance was. Herodotus has a uniquely privileged position in being regarded as the father of history (Herodotus 2014). This comes about from him being the first narrator who tried to systematise the chronology and narrative of history, although much of what he describes is regarded as hearsay and some of it is somewhat fanciful. His only known work *The Histories* is ironically also the earliest Greek work to have survived intact through time.

One of the areas where we know that Herodotus found shells was a valley in Egypt in the Mokattam mountain. This is now a suburb of south-east Cairo, but in the time that Herodotus was visiting, it was an area where limestone had been quarried in significant amounts as large blocks. The primary use of the quarried stone had been to construct the pyramids at Giza, 2000 years before his visit. Sadly, the area is now more of a land-fill site for Cairo. It was at Mokattam that Herodotus saw what he described as backbones and ribs of winged serpents that had been killed by ibises, a bird which he claimed was of considerable use as they killed such creatures as these. What he had seen were almost undoubtedly fossils, as these tended to be washed out of the quarried areas after heavy rain and his explanation was one which fitted in with the belief system which was current at the time. Although we would regard this as mythology, at the time it was a completely believable literal interpretation of observations fitting a preformed notion of the world.

With *The Histories*, we have more than just a complete work by a named and known author, we have a true rendition of what Herodotus saw, thought he saw, and his interpretation of his observations. His interpretation might be strange compared to modern ideas of palaeontology, but nonetheless he describes fossils and poses questions regarding their origin and significance. This is contrary to the problems which many other commentators of the ancient world have left us to unravel. In the case of Xenophanes, for example, since we do not have his original works, our knowledge of his philosophy is influenced by interpretations and nuances given to it by his translators and transcribers throughout the ages. We should, therefore, always be cautious in leaning too heavily on later commentators in explaining what philosophers of the ancient world were really able to infer from what they saw.

The situation tended to change with time as literacy improved and the number of people both reading and transcribing classical texts increased. The spoken tradition of passing on ideas moved towards a written tradition, with books surviving either in their original form or as accurate copies. This change increases the fidelity of knowledge and we can be more certain that ideas attributed to a specific individual have the correct provenance. None the less, the translator's hand can still make inadvertent changes of emphasis. There were, for example, cases where the original Greek versions were kept as Arabic translations, only many centuries later being retranslated into Greek in the versions that we now have.

Later on from *The Histories* of Herodotus, we find more references to fossils, this time from Eratosthenes (276–195 BCE). Known only from collected fragments of his work, he comments that large numbers of shells can be found in Cyrene, North Africa, miles from the coast of what is now Libya. He also, quite correctly, takes this as evidence that the area was once under water (Eratosthenes, Fragments trans. Roller 2010).

Identification of marine shells, a commonly found animal product around the Mediterranean basin, recognisable to any person from the coast, would not have been difficult. Neither would the obvious conclusion be difficult to come to that wherever they were found, it had to have been underwater at some point in the past. Put another way, if the number of shells was so large that it would be implausible for them to have been dumped there, then a simple assumption to be made is that the sea had once covered that area. This in itself is not a significant observation of startling originality, and the thinkers of 2000 years ago had just as clear a perception of the observed world as we do. What would have been significant would be to have worked out by deductive reasoning a cause for the area to have been under water in the first place. It was the Greeks who were the first culture to develop ideas not based on observation to explain observed phenomenon, and this was due mainly to followers of Plato. It would take the most famous follower of Plato, Aristotle, to develop and use observation to fuel-reasoned deductions. This was a skill at which he excelled and was especially well applied to biological phenomenon.

This Aristotelian line of reasoning – creating an explanation of an observed phenomenon based on knowledge and experience – was epitomised by a Greek geographer and philosopher who went by the name of Strabo. He was born in 63 BCE, died in 23 CE, and was part of an affluent family in what is now Turkey. Strabo was taught by Xenarchus (first century BCE) who was broadly Aristotelian in outlook. Nonetheless, Strabo developed his own outlook on the material world, still based on the ideas of Aristotle, but in many ways more wide ranging. Strabo travelled widely and is best known now for his work *Geographica* (Figure 2.2). In this work, Strabo discusses in some detail the conundrum of marine shells being found so far from the sea and frequently at quite considerable altitudes.

Instead of simply jumping to a conclusion, he reviewed other ideas which attempted to address this question, writing down ideas and conclusions. His first review was of Xanthus of Lydia who worked about the middle of the fifth century BCE, like many such classical scholars, only fragments of his work remain. Xanthus pondered finding rocks in the shape of sea shells so far inland and came to the very logical conclusion that the entire Anatolian peninsular was previously under the sea, which had partially dried up to reveal the land beneath. This he compared to the way that rivers or lakes will dry out in times of drought to reveal the lake or river bed. Although most areas of the globe are familiar with seasonal

Figure 2.2   Frontispiece of *Geographica* by Strabo. This edition is in Latin and is from 1620. The original would have been written in Greek, but the only fragment in Greek is from about 500 years after Strabo died.

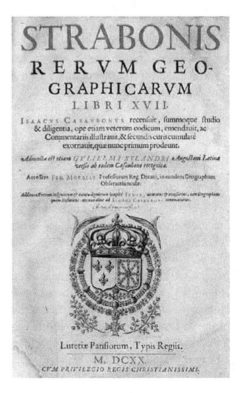

droughts, the Mediterranean basin is sufficiently prone to this that it was easily understood as a practical event and could be accepted as possible. Strabo went on to review the ideas of Strato on the matter of stranded sea shells.

Strato of Lampsacus (335–269 BCE) had attended Aristotle's school, after which he went to Egypt as tutor to Ptolemy II Philadelphus where he developed his already well-tuned interest in natural history. Strato had previously observed that the amount of silt brought down by rivers into the Euxine, the contemporary name for the Black Sea, should have raised the lake bed. Working from this simple observation, he surmised that there may have been ancient lakes which had eventually had their borders breached as silt was accumulated, emptying their contents into the ocean and stranding bivalve shells of the indigenous molluscs well above sea level. Strabo rejected this idea as not being able to account for the widespread distribution of shells (Strabo 1917-1932). What Strabo concluded was that deluges, earthquakes, volcanoes and swelling of the land beneath the sea, all phenomena he was familiar with, could account for the stranding of marine shells so far above the current level of the sea. Strabo also makes reference to the well-known fossils found at the site of the pyramids in Egypt which resemble lentils in shape. These had previously been explained away as remnants of the workmen's food, lentils, converted somehow into stone. This he regarded as very unlikely but did not offer any counter-explanation. We now know these lenticular fossils to be foraminifera which are common in Eocene and Miocene deposits, which correspond to the age of the stone used in construction of the pyramids.

Most of the earliest speculations regarding fossil organisms were based around finding sea shells, mostly marine bivalves, far from the sea and often at high altitude. At this time, that is around 2000 years ago, it was not so much about whether fossils were of animal origin that was vexing philosophers, they had no problem with that, it was how these shells got where they were that was the problem. These shells would have been readily observed as much because of quarrying activity as natural erosion of barren surfaces, exposing them on cliff faces and coastal outcrops.

There developed two lines of reasoning for the presence of these banks of shell containing sedimentary rock. The first was that the sea rose up and receded on a cyclical basis. The second was that the areas in question had been under water but had been lifted up by catastrophic events, such as earthquakes or volcanoes. In a general sense, the idea of a watery planet as suggested by Anaximander (Figure 2.3), Xenophanes, Herodotus and others was a powerful argument because it did explain the observations. It was also given a great deal of plausibility because of the philosophers associated with it. Since it was impossible to go back in time to test the idea directly as to whether the seas could rise up, it was also very difficult to contradict. An argument in favour of a watery planet periodically inundated could be backed up by the known world being relatively small compared

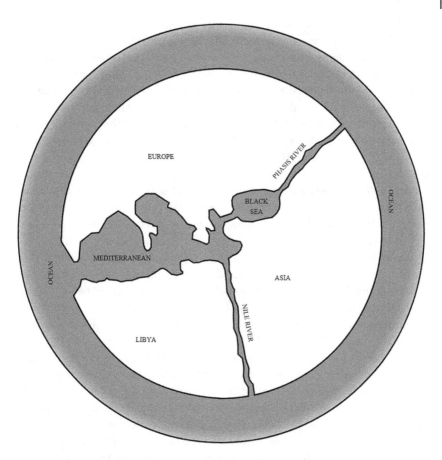

Figure 2.3   A reconstructed image of the world map of Anaximander (611–547 BCE). The habitable land was effectively an island with the surrounding sea capable of inundating the land. The Phasis river is now the Rioni River in Georgia.

with the known seas. There was an implicit understanding of this as the oceans as they were travelled in trade had no end as far as the seafarers of the time knew. It was generally assumed that the world, the dry land, was surrounded by an endless sea. In which case, there would be plenty of water to inundate the land from time to time. Given this disparity between land and sea, it was easy to conceive of a time when all the land was under water for long periods. Certainly long enough for the establishment of great colonies of marine organisms, the shells of which became stranded when the sea receded, or the world dried out.

The ease with which marine shells on land were recognised stopped detailed investigation of one particular question – are they the same types and species as are found in the sea? The assumption seemed to be at the time that they were; they looked like them, so they must be them. During this period, there was a very nebulous form of taxonomy, based around the work of Aristotle. He spent considerable time and effort in both collecting and observing living things and was better informed regarding natural history than any of his contemporaries and successors. Indeed, he maintained this pre-eminent position until relatively recently. He clearly understood that theoretical ideas demand facts before they can be accepted. Without facts any idea is an hypothesis, not a theory. With his wide knowledge of the biological world Aristotle constructed a taxonomy, which in many ways is quite correct, being based on multiple characters, rather than the Platonic method of simple dichotomies. Aristotle considered that species were immutable and complete, meaning that although one species did not evolve from another, overall they could be arranged into a hierarchy from what was considered to be the least developed to the highest and most complicated. He was also very keen on teleology to explain the various features that an organism had, or in his mind was designed for. It was, therefore, assumed that the seashells found high up in cliffs and miles from the coast were representatives of species that could be found in the depths of the oceans if they were looked for hard enough. Assuming they were deep sea species was a good way of avoiding having to explain why they were not found living in shallow seas which were accessible to fishermen.

That there was no systematic attempt to answer the question of whether these fossilised shells were the same species as those already known is hardly surprising. For example, the miocene lenticular fossils of foraminifera found at the pyramids are very small and there were no artificial aids to magnification available to the philosopher scientist of the day. The scale of sizes which could be investigated successfully ranged from the smallest thing the naked eye could see to the most distant star visible with the naked eye, neither microscope nor telescope would be available for almost 2000 years. In a similar way, shells could only be brought up from the depth that a diver could reach on a single breath. In the Mediterranean while pearls and pink coral were valued, it was fishing for food that was the most significant activity, marine taxonomy took last position in fishing activity.

With the sense of immutability of species that was implicit within the taxonomy of Aristotle, there also came a sense of permanence, as there was no concept of extinction. This created some potential problems with found fossils that were obviously not marine molluscs. Larger fossils that were commonly encountered and were apparently part of a large animal could not easily be dismissed. These fossil skeletons had to be incorporated into an understanding of the world that was current at the time. In other circumstances, such fossils would be quite difficult to accommodate within a purely secular society where extinction and immutability were unrecognised. This is as much because of the relative abundance of fossils in the limestone areas of the Mediterranean basin, of relatively recent mammalian origin, made them difficult to ignore. It would have been a case of trying to explain what was unexplainable within the understood science of the day. However, this was not a secular society. The religion was a polytheistic one made up of numerous myths and legends (Bullfinch 1969), which would stand the discoverer of these ancient mammalian bones in good stead.

This situation of myths and legends created an interesting paradox in interpreting some of the fossils that we have evidence the ancients knew about. There is considerable reason to believe there was not only knowledge of ancient bones but also that they were collected and sometimes displayed. These displays were not as animal bones, but as if they were artefacts of ancient civilisations, quite different to their contemporary society.

One good example of incorporating fossil remains of long extinct species into a contemporary narrative, for which a well-argued case has been made, involves the Mediterranean Proboscidea, the order which contains the elephants and mammoths. This order was once represented by a range of species across the whole of the Mediterranean basin, as well as further afield. The species which were found on the islands of the Mediterranean seem to be dwarf species related to the much larger elephant species we have remaining in the tropics and the even larger mammoth species of the larger land masses to the North.

Through geological time, there does seem to have been a significant process of allopatric speciation amongst the now-extinct dwarf elephants of the Mediterranean islands (Young 1969). These islands along with the mainland around the Mediterranean have a geology which has resulted in most of the large fossils being mammalian. Of these, to a large extent, elephants make up a significant amount.

One of the elephant species which is thought to be of particular note is *Palaeoloxodon falconeri*. This was a species which was less than a metre tall at the shoulder and weighed about 300 kg. It probably arrived on the islands when the sea was about 100 m lower, during the Pleistocene, and land bridges were available for the animals to cross between islands. There would have been stretches of water between some of the nascent islands with fast running currents between them, even so, these would appear to have been navigable. It was *P. falconeri* and closely related species which Othenio Abel suggested in 1914 to have either originated or encouraged the idea of cyclops. This interpretation was based on finding fossil elephant skeletons washed from coastal areas and river banks with their skulls. The skull is important here because the high position of the nasal opening in elephants is singular and central, and therefore easily mistaken for a single eye socket (Figure 2.4). As the bones of the skeleton would have been disconnected, reassembling them in an upright pose with a cyclops skull on top would not have been seen as an unrealistic reconstruction. We now think of cyclops as being part of the mythology and folklore of the area, but at time they were considered to be a real part of the historical landscape.

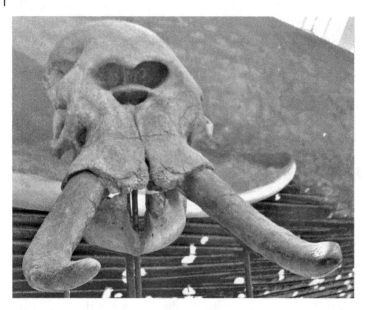

Figure 2.4   The skull of a small species of extinct elephant, the Pleistocene species *Cuvieronius platensis*. This particular species was from South America but shows the nasal openings which could be mistaken for an eye socket.

Some of the skeletons which would have been found would not have been particularly old in geological terms. For example, *Palaeoloxodon cypriotes,* sometimes called the Cypriot dwarf elephant, was estimated to only weigh about 200 kg. *P. cypriotes* was thought to have become extinct approximately 11 000 BCE, yet an even more recent extinction was *Elephas tiliensis* from the small island of Tilos situated halfway between Rhodes and Kos. This small elephant species seems to have held on as the Mediterranean basin became populated, finally becoming extinct about 4000 BCE, which is clearly within the period of historical human activity. There is little to indicate that these species became extinct through human activity, but it has been suggested that in any given geographical area the arrival of humans is indicated by the disappearance of the megafauna. This is a very simple explanation, but one which does frequently hold good.

When the skeletons of extinct species were first discovered, it was recognised that they were obviously animals, there was no attempt to pretend that they were some other product of the earth. This was a problem that would arise with the monotheistic religions much later. These animal remains then had to be put into a framework of zoology as it was recognised at the time. The question of origin of these fossils, either from extinct animals, or coincidental creations, would not arise as a serious consideration in a culture of myths and legends. This would change with the monotheistic Christian church. This belief system assumed absolute truth as part of their dogma, making it impossible to imply fallibility in God by suggesting an animal may have been created imperfect and therefore capable of extinction. There was a major difference in ancient Greek theology as it was broadly polytheistic, there were many gods and lesser supernatural beings who frequently interacted directly with mortal man. It also had a remarkable tolerance to a range of ideas, for example, Plato (c. 429–348 BCE) considered reincarnation to be a real aspect of life and death, whereas Epicurus (341–270 BCE) denied any form of afterlife.

The Romans also had a polytheistic system with a number of deities which were greatly influenced by the Greek system of religion. This was most probably because of the large number of Greeks living on the Italian peninsular. What this broadly meant was that any bones found were going to be shoe-horned into their complicated theology, if they were obviously not from animals that were known. This rationalisation of ideas is demonstrated by the work of Palaephatus. Almost nothing is known of this author, even his works are only known from later fragments, but it would seem he was writing about the middle of the fourth century BCE. Palaephatus wrote down a series of Greek myths as they were known at the time. He then described the sorts of perfectly normal events which could have given rise to the myth based on continual repetition and embellishment. In these explanations, he does not mention the gods widely, most probably so as not to give offence in a deeply religious period of Greek history.

At the time of the Greeks, the Romans and other Mediterranean states, there was no contemporary knowledge that elephants had ever been living in the Mediterranean basin. So for the discoverers of these skeletons of Elephantidae, the question arose as to what they were. One idea which has been given much thought is that elephant skeletons could have been interpreted as skeletons of cyclops (Abel 1914). We cannot be sure whether the idea of a cyclops came before the discovery of the elephant skeletons, or afterwards, and in many ways this is irrelevant. What we can have some certainty about is that these discoveries reinforced the idea of a race of one-eyed giants, giving a verisimilitude of authenticity to the otherwise mythological status of the cyclops. It is not always worth pursuing the line that myths have a basis in reality for their construction. The unicorn is such a case, it is just as likely that the unicorn came from a fertile imagination, as being based on an observed animal. It can always be demonstrated that to create a mythical beast, the imagination has enough fertile ground in itself for their construction. Contemporary science fiction aliens attest to this. But where there is a confluence of ideas and fossil remains that echo each other, it is reasonable to assume that the manufactured beast could attain the status of a real animal in the public perception.

Within the terrestrial mammals, an elephant's skull does have some notable features which separate it from most other species. It certainly separates it from all of the domestic and hunted species that would have been found at the time. All of the mammals associated with man, from commensal species such as rats and mice, through domesticated farm animals to the hunted boar and deer, the skull is constructed in broadly the same way. The snout is elongated with the nasal bone and consequently the nose pushed far forward. The orbits of the eyes being much further back from the nose. With this common plan, any skull likely to have been seen, whether from a dead animal or one for the pot, would be recognisably animal. On the other hand, an elephant skull is rather different. Even the dentition is significantly different, there being no canines and the incisors elongated into tusks. This is not the only adaption that is found in elephants and neither is it the most significant. The skull is foreshortened and the small nasal bone is high on the skull, with the orbits of the eyes, which are not clearly defined, below the nasal opening. So from the skull, the nose appears to be on the forehead, although as we know, the nose and upper lip have become the trunk. Therefore, an elephant skull will have the appearance of a massive malformed human skull with a single orbit in the midline. It would certainly look more like a human skull than any of the skulls which would have been known from the animals with which they were familiar.

If the ancients believed that cyclops were real and the skulls of extinct species of elephants were the skulls of dead cyclops, then the associated skeletons should be the skeletons of the cyclops as well. It would, as a simple consequence, only be necessary to rearrange the bones as if from a standing, bipedal, species to complete the picture. Such a situation of having a skeleton as well as a skull would have been a rare one indeed. It is quite likely that from time to time, enough bones were washed out of cliff faces to warrant laying out a partial reconstruction, if only to satisfy local curiosity.

In Greek, cyclops literally translates as 'round, or circle, eyed', and this is the word we habitually give to stories of monocular humans and animals. Although we use a Greek word, the story of cyclops is not confined to Greek mythology. These stories appear in many other cultures. Not so far from the eastern end of the Mediterranean and the northern coast of the Black Sea, the Scythians had the Arismaspi, a one-eyed people who stole gold from Griffins. As we shall see later, Griffins are another mythical species that may have originated or evolved from fossils discovered in antiquity. Again, these would have to be explained by the local population as they obviously represented an animal that had once been alive. In the Philippines, there were folk stories of a giant cyclops which permanently giggles, called Bungisngis. The one-eyed monster of Japanese folklore is called Hitotsume-kozo and in Albania there is a man-eating cyclops by the name of Katallan. There are many other cyclops myths from around the world. It would be disingenuous to claim they were all a result of, or even encouraged, by fossil elephants, but some may have been. It should be self-evident that not all of these myths would have had a recognisable origin associated with elephant remains. While most, if not all, of the population of the northern Mediterranean would never have seen the skull of a contemporary elephant, this would not be so elsewhere. Many eastern and southern areas had indigenous elephant species and therefore a familiarity with the skeletal structure of these animals would have been relatively common, especially where they were hunted for ivory.

Mention has briefly been made, above, of another mythical beast, Griffin. This would be traditionally described as having the body and legs of a lion and the head and wings of an eagle. Sometimes the front feet were depicted as talons, but this is of no palaeontological relevance. It has been suggested by Mayor (2000) in a detailed analysis that the Griffin was not so much an imaginary species as one that grew out of attempts to put found material, in the form of fossils, into a human context. This would be in much the same way in which it is suggested that elephants gave rise to the myth of cyclops.

Based on illustrations from classical documents and stories of mythology, a very plausible argument can be made that the Griffin, as originally portrayed, was in fact the product of imagination. The idea pivots around a created reality formed from the fossilised skulls of dinosaurs of the suborder Ceratopsia. The ceratopsids as a whole were herbivorous dinosaurs broadly categorised by having a beak-like mandible. Most of the known species seem to have also had a bony frill at the back of the head with some also having horns facing forwards. As a group they seem to have originated quite late in dinosaur evolution, appearing in the second half of the Cretaceous Period. Not only do they seem to have appeared quite late on, but they also had a wide range of sizes from small species up to massive species like *Triceratops* (Figures 2.5 and 2.6). They seem to have originated in what is now Asia, radiating very quickly to become widespread throughout the Northern Hemisphere.

Figure 2.5 Skull of an American species of *Triceratops* in the Birmingham Think Tank.

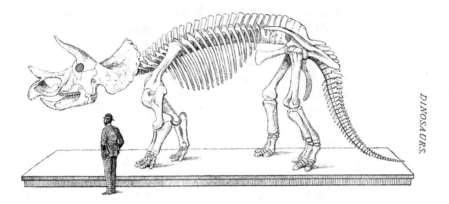

DINOSAURS

Figure 2.6 *Triceratops prorsus,* illustrated with a man for scale. *Source:* From Hutchinson, H.N. (1892) *Extinct Monsters.*

The bulk of ceratopsid species have been described from specimens originating in North America, and from areas in Mongolia. They are also found at sites nearer to the Greek sphere of influence on trade routes, such as Kazakhstan and Kyrgyzstan (*Times Atlas of World History* 1979). It seems that ceratopsids lived in large herds, and they were certainly abundant as there are some very large fossil beds with large numbers of skeletal remains of individuals, both in Asia and North America.

The distance from the Mediterranean basin to the Asian fossil beds is important, because the skeletons were not, so far as we know, ever seen by the Greeks or earlier civilisations, such as the Egyptians. The legend, if it did originate with a fossil find, came from much further east where the geology not only allowed for the survival of these remains but also had gold deposits. It was the presence of gold that created an ancient tradition of gold mining activities. Originating in mining activities, the suggestion is that uncovered remains resulted in stories of Griffins hoarding, or guarding, mounds of gold. If the mythical beast did originate with a fossil, then it would most likely be species from one of two genera of ceratopsid, either *Protoceratops* or *Psittacosaur*. Both of these were beaked, but there are differences in that while *Protoceratops* was beaked it also had a bony outgrowth, a frill, which lay behind the head. *Psittacosaur* differed in not having the frill, and this is a genus which has about 10 described species.

It is easy to see just from their size and appearance how these beaked herbivores could, from skeletal remains alone, be raised up as monsters. Being so unlike anything which would be encountered by the people of the area where the skulls were found, some explanation of the remains would have been needed. It does seem possible that this was either an originator of the Griffin myth or a significant reinforcer of the story. It is also reasonable to suggest that since this would only be an oral description, embellishment and adaption would be expected over time and distance as stories were retold and images copied and recopied by hand. This alteration over time could quite easily explain the provision of wings on Griffins, for example, since true wings have never been present on a four legged animal.

Suggesting that marine shells in high mountains and very great distances from the sea indicated that the area had at one time been underwater was not controversial. What was a source of debate was the method by which it happened. These fossil shells were, after all, quite obviously the same in broad structure as extant forms. This was not so with the skeletons of extinct terrestrial species which turned up in rock formations from time to time, washed out of cliffs or found in quarries. As far as we know these large fossils were not dug for specifically, they were chance finds associated with landslips and storm damage. They were obviously skeletons, so they had to be fitted into the knowledge of life as it was perceived at the time.

For the philosophers of the classical period, fitting fossils into any taxonomic system would undoubtedly have caused some problems. At the time, skeletal remains were not enough to fit a species into a taxonomy. All the features normally used were based on the external phenotype, such as fur and feathers, scales and fins. It should be remembered that while skeletal remains may be found, it is exceedingly rare to find a whole skeleton. For example, as recently as 2008, a Pliosaur was found in Lincolnshire in the UK which was described as a substantial find containing a large part of the skeleton. What was actually found was a tooth, 29 vertebrae, 14 ribs and a number of limb bones. No doubt this disruption and loss to the skeleton was due to contemporary scavenging when the animal died, as well as earth movements, but it does serve to illustrate the incomplete nature of most large animal fossil finds.

Besides the fundamental problem of not being able to slot a fossil into a zoology that did not contain the concept of extinction, there were other problems. One such was trying to fit found remains into existing schedules of life, when they had no comparable contemporaries. But there was a bigger problem, one associated with the positioning of fossils within strata. Fossils of what appeared to be large animals, sometimes they were clearly recognisable as such, were found at different levels in the ground, some even having been found by miners. How did these remains, if that was what they were, become buried so deeply? In the sixth century BCE, Xenophanes described fossil fish that had been found in mines on Paros, for example. This conundrum was not going to be resolved for a very long time.

There is no doubt that references to fossils discovered around the Mediterranean can be found quite widely in classical literature, the philosophers of the day being familiar with both stranded sea shells and larger fossils. They would on occasion relate the level of consternation that local miners and quarrymen would express at the scale of the monsters that they had unearthed. This is quite understandable as they would have found some very large mammalian fossils, although due to the geology there would be no dinosaur remains in the immediate area (Biju-Duval et al. 1974).

This is probably why stories of Griffins would have come from further afield than the Mediterranean, where the more remarkably shaped reptilian fossils would be found. Since the reptilian skeletons from the East would not have any living descendants to which they could be related, imagination would have fleshed out the skulls in terms that could be related to known animals. The chimaera that is the Griffin could be explained as a simple product of the imagination (Wyatt 2009) constructed from religious imagery and symbolism. This, like all arguments, sets up unnecessary lines of conflict. The story is essentially impossible to dissect from the past but is most likely to have many elements from each side, mythology and religion, fitting details from many stories together into a composite whole. The problem with this particular story is that it crosses across many disciplines, each of which wish to be in the ascendency of the argument as to its true origin.

This contrasts with the larger fossils of the Mediterranean, most of these being mammalian. When they were found, they could be directly compared with both contemporary animals and human remains. The obvious thing, then, when faced with enormous bones is to describe the animals of the past as simultaneously giant, which they were, and beasts of food and burden, which they were not.

By implication, large beasts of burden require large people to tend them, and so it was that the myth could have arisen that in the past, the islands of ancient Greece were the home to a race of giants. The occasional unearthing of such remains would serve to reinforce this notion. It is certain that contemporary chroniclers of Greece would frequently bemoan the reduction in stature of mankind. In this respect, Pliny the Elder was no exception (J. Healey (trans.) 1991). Although he was Roman, rather than Greek, he was writing in the first century CE and commented that it was a matter of common observation that the entire human race was smaller than in past times. Interestingly, he also quotes Homer, who would have penned his epic works around 900 years earlier, as also observing that humanity was getting physically smaller. Homer's observation would also be a reflection of seeing large bones which could only have come from giants of antiquity, being mistaken for the normal size of individuals.

In some works, it was said that the very large fossil skeletons, which it should be remembered would not have been complete or with the bones in the right order, were those of giants. By contemporary presumption, these were said to be a result of the Gigantomachy. This was a battle in which the giants were fighting the Gods of Olympia for control of the Universe as it was known at the time. The nature of the battle was sufficient to explain the large skeletons in the minds of literal believers or to confirm a belief in sceptical thinkers, because it seemed less likely that these remains were animal in origin than remains of giants of the past.

It is interesting to observe in this context that reported sizes with regard to supposed human remains varied enormously. These ranged from very big, but plausible, to enormous and beyond any possible size for a human, such as over 15 m and more in height. These are only estimates as there will always be two problems with strict comparison of units. The first is that ancient units may be spoken of with authority, but their size is unknown. The second is that the units as described lacked the precision of modern SI units. Put another way, while a hand span may be routinely used for small items, the variation in size as measured would make it useless for distances. In more practical terms, for example, in the Bible, Goliath is described as being 'six cubits and a span' in height which puts him at approximately between 2.89–3.35 m. The range here is due to the variation in interpreting what these ancient measures were, as there were no standard units or values for them. To put this height into a modern context, the tallest recorded living man was said to be 2.71 m in height, and he died at the age of 22 years due to complications associated with this unrestrained growth.

One of the intriguing aspects of the attitude to fossil discoveries found in classical literature is the apparent disparity between the popular belief and the reasoned philosophical approach to them. While the artisan discoverers of large bones and skeletal remnants might be inclined to put them into their existing framework of myths and reality, the philosophically inclined might not. They would question their origins, and exactly how they came to be where they were found. For those that discovered such fossil remains, there were no such questions. The bones were there and fitted into their belief system with little effort of interpretation. This is bearing in mind that just as some take the Bible literally, so some would have taken what we would now describe as myths as simple descriptions of historical fact.

Although there is some contention in the suggestion, it is possible that we can see this in the remaining works of Empedocles of Acragas (1998). Empedocles was working around the middle of the fifth century BCE and constructed a strange collection of biographical details about himself, such as that he was a God and had previously been a bush and a bird. His work has come down to us only in fragments and was, so far as we know, written entirely in poem form. This makes it is difficult to be precise about his meaning, but he was well considered and discussed by Plato and Aristotle. It is mainly from these two sources that we gain most of our knowledge of the ideas of Empedocles. In his poem *On Nature*, he describes a complicated process of construction and disintegration of organisms based on four elements of fire, air, earth and water and two opposing forces of love and strife. Aristotle claims that Empedocles was the first to make a clear statement of these four elements.

Empedocles claims that there was a time when separate limbs wandered around on their own and faces without necks and eyes 'in need of foreheads'. It is these disembodied parts which are used by Empedocles as examples of strife disintegrating bodies and of him recognising fossilised material as coming from ancient organisms of a different type to the mythological bones of dead heroes. He also implied that random body parts could come together to form inappropriate organisms that were incapable of survival, and these, too, would disintegrate, returning to the ground. It was only when the correct parts had joined to form complete organisms which could then go on to reproduce, replicating themselves in perpetuity, that a true species was formed. One of the examples which Empedocles used was that of the half-man half-horse Centaur, which he regarded as untenable organisms, which is why they no longer survive. This concept of species being of mixed origin was going to be highly contentious, not least of all because it introduced an element of randomness into what was increasingly seen as a perfect cosmology.

By the fourth century BCE, Aristotle had taken against the idea of Empedocles that species were more or less constructed randomly from mixed parts. The Aristotelian arguments for this were based on physiological considerations. Looking further into his ideas of zoology, we can recognise that he had a clear idea of the nature of species, although they were not necessarily expressed as clearly as a modern taxonomist would wish. Aristotle was aware of the Platonic view that species were fixed, immutable and eternal, but he modified this in such a manner as to allow for the variation we see between individuals in distinct species. His ideas were based around three concepts.

- All the visible signs of an animal type have to be present. So for example, all those things which would describe a fish should describe all fish.
- There are indivisible forms within the group which describes specific types, that way we can recognise a trout as being a trout.
- The form is controlled by knowledge and method so that a child has the broad appearance of an adult. The two are the material cause and the efficient cause.

This last factor does rather fall down when looking at larval lives.

These concepts automatically precluded the ideas of Empedocles that there could be random production of monsters, whether they survived or not. This idea of saltational evolution associated with macromutations would reappear in a different guise in the nineteenth century with what came to be known as Goldschmidt's hopeful monsters (Goldschmidt 1940).

What the ideas of Aristotle does allow is for a static idea of species that do not alter over time. This small sentence was to become a dogma within the early Christian church at a time when easily stated dogmatic statements would serve as anchors in turbulent times. Aristotle created a system unlike the ideas of Empedocles. While not explicitly precluding extinction, it did cause something of a conundrum when it came to explaining the large fossil bones that were regularly being found within the Mediterranean basin. The paradox is easily stated in that the large bones could not belong to extinct species because there was no extinction; at the same, there did not appear to be any extant equivalent species to be found. It is considered possible that this is why among the known writings of Aristotle no mention of large fossil bones is made. It is always possible that this idea is misleading as Aristotle produced a considerable catalogue of species and like all taxonomies this starts with species that are living, and it is much more difficult to construct a taxonomy based only on fragmentary skeletons than it is using a whole animal.

This early classical period of interest in fossilised remains was an unusual gap in the general lack of investigation and curiosity that came before and after this fertile period of intellectual flowering. There are many reasons that this change took place, and some of them to do with social conditions and some to do with the economic situation at the turn of the millennium BCE/CE. This hiatus in intellectual activity was not confined to palaeontology but occurred throughout the intellectual cannon that was developing in the ancient world. The changes which caused this hiccup in the otherwise smooth process of developing ideas were inextricably linked with the social changes inevitably associated with nations whose only apparent motivation was military power alone rather than the greater development of the wider population.

There is another aspect to the development of ideas around fossils which is of particular interest. We can understand that explanations can be formulated regarding how fossil shells found on high cliffs got there, but there was no explanation or speculation as to exactly what these shells were. We now realise that such remains, however much they appear to be contemporary species, are long extinct forms, but that is with the application of modern techniques of taxonomy. Originally, no attempt was made to treat them as different species to those which were known. If they were so different, they had to be recognised as unknown, and it was just as reasonable to assume them to be alive somewhere, but not yet found. Once a view of the world was created that included some notion of the position that fossils held, this would be a stable paradigm by virtue of the authority with which it was promulgated and kept at the front of any discussion of natural history or palaeontology.

Attempts to explain the passage and development of scientific progress in terms of the way that changes appear. The philosophical notion of ideas achieving their right place and time has been made, some with considerable apparent success. The work of Thomas Kuhn is one such which is worth particular note. In his publication *Structure of Scientific Revolutions*, which first appeared in 1962 (3rd ed. 1996), he makes a very reasonable case for the way in which science can be seen as moving forward. His hypothesis for scientific progress was based on a model which seems counterintuitive. This is that instead of a slow and continuous progression from one small discovery to the next, occasionally punctuated by the brilliant thinking of a few individuals who shifted the emphasis from its current path to another, Kuhn postulated a punctuated equilibrium model.

For those readers of a specifically biological bent, the term punctuated equilibrium will have a very precise meaning. In this context, however, it refers to the ideas of science becoming embedded with small advancements being made until a revolution takes place that changes the scientific view of the world. This quiet process of investigation is considered to accumulate discrepancies in data which do not fit the current model being used. These data outliers are ignored until such times as they are impossible to work around. These will then strain the science within which they operate until something has to give. Which is usually the model that is currently being promulgated as fact.

This very reductionist idea of the progress of science works well in some circumstances, for example, the motion of the planets around the sun. We assume our model to be correct because it has no gaps of prediction, but for many centuries neither did the geocentric model of the solar system. It was really only with the progressive complexity coming from more accurate methods of measurement, especially associated with invention of the telescope (Wall 2018), that it came to be realised that as a model of reality it was starting to fail. In this particular case of course, it was the non-scientific vested interests of the Church which held up progress towards a more sustainable model of the Universe.

Although this method of seeing scientific advancement as being associated with accumulation of nonconformist data until it reaches an activation energy, it works in some cases. For other subjects, like biology, where data can have a very large normal range, it becomes a reductionist and simple view of a very complicated subject. Other factors which cannot be overlooked include technology. Before the advent of the microscope, there could be no concept of a cellular basis of life and no concept of substructure within an organism. Similarly, the telescope rendered so much more visible in the night sky that all of the objects that were found were incapable of being fitted into the old cosmologies. More recently the Polymerase Chain Reaction (Saiki et al. 1985), for which Kary Mullis received the Nobel Prize for Chemistry in 1993, is an apparently simple method of replicating DNA artificially. Nonetheless, it now underpins most if not all genetic diagnostic tools and forensic DNA investigations as well as sequencing techniques. These are technologies without which some of the questions which have since been answered could not have been asked in the first place.

New technology and paradigm shifts due to accumulated discrepancies apart, and it takes an economic determination on the part of society to pursue science in any form. Put another way, without both time and space, science cannot progress. The implications for palaeontology in this context are clear. For centuries, the old working model propounded by Aristotle and his followers did not allow for the massive skeletons to be seen as significant, quite possibly they were tacitly ignored from their taxonomies because they belonged to mythology and were therefore beyond worldly investigation. Of course, they may have been ignored because it was, at the time, impossible to identify which animals they were. For whatever the reason, the situation was going to persist for many centuries CE as the social and economic fabric became less able to sustain purely intellectual activities. Even when the situation eased and the established Church became a wealthy and powerful player, the vested interests kept a very tight hold on intellectual activities, curtailing anything that might question the literal belief in the Bible or the Aristotelean view of nature.

During the early years of the Church, control of formal education was one of the best ways of influencing society and keeping the state under control. It was the study of subjects mired in archaic ideas that encouraged biblical scholars to investigate the age of the planet, but based strictly on literal interpretations of the Bible. These estimates are based on a fundamental misunderstanding, but what they do show is the extent that written records can survive. What they do not show is the age of the earth, while they do give us a good indication of the antiquity of civilisations around the Mediterranean basin. Many of the estimates of the age of the earth are based on tracing ancestral events in the Bible and this is where the fundamental misunderstanding appears, and the age of humanity with a written record and the age of the earth have no common point of contact.

Most of the earliest guesses of the age of the earth based on religious texts gave a date for creation of about c. 5500 BCE, and these attempts being based on details of the Septuagint, substantially the Old Testament translated from Hebrew to Greek, or later on the Masoretic text which is the Hebrew and Aramaic texts of the Tanakh. These were not the only methods used to guess the age of the planet, but all that were delivered with a sense of certainty were taken seriously in their own time.

In 703 CE, the Venerable Bede came up with a different estimate, which was also much more precise than most of the previous ones. He suggested creation started on 18 March 3952 BCE, this was considered heretical by the Church of the day as it was in contradiction of all previous estimates. Guessing the age of the planet was not even confined to the overtly religious, and Isaac Newton thought the creation was about 4000 BCE and Johannes Kepler came down in favour of a date of 3977 BCE. Of all the attempts to date, the planet based on ancient writings probably the best known estimate of the age of the earth was suggested by Archbishop Ussher, Archbishgop of Armagh from 1625 to 1656. His work was published in 1650 and 1654 where he produced a chronology that worked well given a literal belief in the Bible. By tracing biblical events backwards through time, he arrived at 22 October 4004 BCE. This was regarded as accurate and reliable and is still thought to be the most influential of the various different estimates.

What all these have in common, besides their fallacious basis, was that no account of the observed world was made during the calculation. Similarly, there was no acknowledgement of the existence of fossils, that discomforting aspect of biology. Extinctions were not necessary to assess since the idea of extinction was not accepted as real, so there was no question to be looked at. By having these limitations, there was a considerable brake on the development of ideas regarding what fossils represented. For this reason, there was a long period of time when the perceived problem of fossils does not seem to have been addressed. This should also be tempered by remembering that at the time and in most parts of Europe, fossils were rare finds indeed and most fossils are not complete bones, but parts of bones. This literal dependence on the Bible for all aspects of ancient history beyond the classical period went on for centuries into the first millennium CE.

By the time of the sixteenth and seventeenth centuries, ideas had changed considerably. Exploration had broadened the horizons both literally and metaphorically, while trade had increased the wealth of Europe. Increasing wealth brought with it an increasing ability to examine subjects previously regarded as not worth pursuing. One such was geology, fossil fuels in the form of coal had been collected and partially dug out since Roman times, but now larger and better organised mines were being sunk. This activity required development of some knowledge of where best to start digging and this meant geology.

To gain a better understanding of why it was that the otherwise arcane study of fossilised material moved from obscurity to centre stage, it is worth considering the social changes that were taking place in this period. During the sixteenth and seventeenth centuries, the social changes presaged a period of increased curiosity about the natural world; questions were asked that had not been thought of before. Changes in the intellectual climate can be summed up by the Church of England breaking ties with the Vatican; it was no longer slavishly followed as an authority. In this period, the world became familiar with logarithms, telescopes and microscopes, gravity and the laws of motion. In 1660, the Royal Society was founded in London. It was in this climate of investigation that new attitudes were bound to arise, and some fossils being seen as self-explanatory, while larger fossils posed intriguing questions that piqued the curiosity of naturalists and required specific explanations.

References

Abel, O. (1914). *Die Tiere der Vorwelt*. Berlin: Teubner. (In German).

Barraclough, G. (ed.) (1979). *Times Atlas of World History*. London: Times Books Ltd.

Biju-Duval, B., Letouzey, J., Montadert, L. et al. (1974). Geology of the Mediterranean Sea basins. In: *The Geology of Continental Margins* (eds. C.A. Burk and C.L. Drake). Berlin, Heidelberg: Springer.

Bullfinch, T. (1969). *Bullfinch's Mythology*. London: Hamlyn Publishing.

Empedocles of Acragas (1998). *Empedocles: Extant Fragments* (ed. M.R. Wright). London: Bristol Classical Press, Bloomsbury Publishing.

Gesner, C. (1565). *De Rerum fossilium, Lapidium et Gemmarum maxime, figuris et similitudinibus Liber*. Zurich: Tiguri.

Goldschmidt, R. (1940). *The Material Basis of Evolution*. USA: Yale University Press.

Healey, J. (trans.) (1991). *Natural Histories: A Selection*. Pliny The Elder. London: Penguin Classics, Penguin Publishing.

Herodotus (2014). *The Histories*. Penguin Publishing (trans. Tom Holland).

Hutchinson, H.N. (1892). *Extinct Monsters*. New York: D. Appleton and Company.

Kirahan, R.D. (2010). *Philosophy Before Socrates*, 2e. Indianapolis, USA: Hackett Publishing.

Kuhn, T. (1996). *Structure of Scientific Revolutions*, 3e. USA: University of Chicago Press.

Mayor, A. (2000). *The First Fossil Hunters*. Princeton, USA: Princeton University Press.

Roller, D.W. (2010) (trans). Eratosthenes). *Geography*. Princeton University Press.

Saiki, R.K., Scharf, S., Faloona, F. et al. (1985). Enzymatic amplification of the beta-globin genomic sequences and restrictionsite analysis for diagnosis of sickle cell anaemia. *Science* 230: 1350–1354.

Strabo (1917–1932). *The Geography of Strabo*. (Ed H.L. Jones) (trans. H.L. Jones and J.R.S. Sterrett) William Heineman, 8 Volumes.

Urmson, J.O. (ed.) (1960). *Western Philosophy and Philosophers*. Hutchinson of London, Publishers.

Wall, W.J. (2018). *A History of Optical Telescopes in Astronomy*. Switzerland: Springer Nature.

Wyatt, N. (2009). Grasping the griffin in Egyptian and west Semitic tradition. *Journal of Ancient Egyptian Interconnections* 1 (1): 29–36.

Young, J.Z. (1969). *The Life of Vertebrates*. Oxford: University Press.

3

The Unfolding Understanding of Fossils

By the time of Shakespeare, around the turn of the sixteenth and seventeenth centuries, attitudes to the natural world were also turning, from slavish obedience to works of the ancients and the Bible, to enquiry as a method of finding the truth. Prior to this, while passing references were made in ancient texts to unearthed bones of very large animals, often taken as being remnants of mythological creatures, or giants from the past, there are no known contemporary descriptive illustrations of the fossils that were found in the Mediterranean basin. This is why we had to wait more than a thousand years until society was ready for the social turmoil of questioning the Bible, and for individuals to appear prepared to pose difficult questions of the accepted geological dogma.

The modern literal interpretation of the bible by creationists is different to the mediaeval interpretation in that most modern scholars think that until the seventeenth century, ecclesiastical scholars were aware of the debates regarding the age of the planet in which the Greek and Roman philosophers engaged. The literal interpretation of Genesis and the Bible seems to date from the Protestant Reformation in the seventeenth century. Initial attempts to reconcile the bible and the new reductionism of Newton and Copernicus were doomed to failure, the measured observations of reality conflicting with the earlier geocentric model of the universe. Attempts to apply the same system of logic to the natural world were disadvantaged from the start because of the fundamental complexity of biology not being so easily explained, and the simplicity of the bible not being so easily believed. It was during this period that enquiring minds started investigating fossils in all their forms, while simultaneously trying to fit findings to previously held assumptions. The outcomes of these reconciliations were often going to be extravagant.

In 1565, Conrad Gesner (Figure 3.1), a Swiss naturalist and physician, published *De Rerum fossilium, lapidum et gemmarum maxime, On Fossil Objects* (1565). This was an important book for many reasons, but primarily because it

Investigating Fossils: A History of Palaeontology, First Edition. Wilson J. Wall.
© 2021 John Wiley & Sons Ltd. Published 2021 by John Wiley & Sons Ltd.

Conrad Gesner par Tobias Stimmer.
(Musée Allerheiligen à Schaffhouse.)

Figure 3.1   A portrait of Conrad Gesner (1516–1565) by Tobias Stimmer of about 1564.

opened up discussion to a wider audience about the organic origins of fossils as we would now recognise them. At the time Gesner was living, this was not a commonly approached question as 'fossil' described any found object associated with the ground. The audience was very definitely an educated one capable of reading Latin, also one with a considerable disposable income as books of this sort were expensive items. While literacy rates were rising during this period, at the time Gesner was working the usual print run for a work was between 400 and 3000 copies of each book and it would only reach the higher number if the initial sales were good enough to justify it. Although this does not compare unfavourably with current practice, the reading public were not overwhelmed with choice and virtually all books printed would be sold. We do not know the size of the print run of *De Rerum fossilium,* what we do know is that the reader would have had a classical education, the book is written in Latin, for an international audience. It also contains frequent lapses into Greek, in the ancient script, while at the same time some of the illustrations have bilingual captions in Latin and German. The acknowledged value of Gesner's work is reflected in there reportedly being about six copies, mostly bought at publication, in the libraries of Cambridge University (Rudwick 1972). One of the important aspects of this book was that as far as we can tell it was the first time a written description of a fossil was accompanied by an illustration. While the Greek and Roman philosophers had pondered the questions posed by fossils, this was always in written form and never illustrated. Later on it was possible to find woodcuts of fossils, but these were not accompanied by any considerable text. At the other extreme Georg Bauer, sometimes known by his literary name of Agricola, published in 1546 *De Natura Fossilium,* On the Nature of Fossils, which is primarily a book of mineralogy (Agricola 1556). In this work, like so many much older texts, the description of fossils was confined to a written one. This can make it very difficult for the reader to know with any certainty what precisely is being described. This problem has a modern counterpart in the use of dichotomous keys for identifying species. It takes a great deal of skill to produce a written key for identification and a lot of practice to make the most of one. This is for the simple reason that it is very difficult to describe unequivocally and with any accuracy a three-dimensional object using words alone. Agricola also produced *De Re Metallica,* which was published in 1556, a year after his death. This particular volume was translated in 1912 by Herbert Hoover, a mining engineer at the time, who became president of the USA.

Gesner was looking at dug-up finds of all sorts, many of which were described and illustrated with woodcuts, a technique that was soon to be superseded by methods of engraving that gives a higher resolution image. Both Gesner and Agricola were wide ranging in their understanding of the term 'fossil' and included gallstones and pearls in their broad and poorly defined usage. Even with this vague idea of 'fossil' as not just dug up, but found in general, it should not be forgotten that in this new arena of found objects, it was very difficult to be sure whether what had been found was organic in origin, or simply resembling an organic object by chance. In some cases what we now recognise as fossils were thought of as natural formations, belemnites, for example, so resembled stalactites that they were regarded as variations on these purely geological features. Gesner was a naturalist with a wide ranging interest, but especially in marine life, which led him to make a significant link between *glossoptera* and shark teeth. We now know that 'tongue-stones', *glossoptera,* are fossil shark teeth, but it took an astute observer and detailed familiarity with living sharks to realise the similarity in appearance was more than just coincidence (Figure 3.2). Like many of his contemporaries, for Gesner most fossils that resembled living forms were simply that, resemblances with little likelihood of actually being the item itself in a different form, transformed by chemistry and time.

As far as we can tell, the first illustration of a dinosaur bone was published in 1677, although it was not recognised as such at the time. The illustration and accompanying explanation of the find appeared in *The Natural History of Oxfordshire* by Robert Plot (Figure 3.3).

Figure 3.2 In 1558, Gesner published Historiae Animalium Liber iv Qui est de Piscium and Aquatilium animantium natura in which he depicts a glossoptera (the separate tooth) and the shark with which he was comparing the fossil. This is the first known comparison between a fossil and a living organism (Gesner 1558).

Perhaps because of the local nature of the subject, this book of natural history and architecture was unusual in being written in English, rather than Latin. Plot was a graduate of Oxford University and the first keeper of the Ashmolean Museum where he was reputed to be a very good curator, producing detailed and essentially modern catalogues of the contents. In 1690, Plot resigned his academic duties citing the lack of funds as the cause. The work of Robert Plot gives an insight into the complexities of thought that were involved in trying to reconcile modern thinking with an entrenched religious attitude. This was not the same as a religious dogma bolstered by a powerful church and vengeful clergy. This was the system that had ruled so determinedly against Galileo and sentenced him to house arrest in 1633 until his death in 1642 for publication of *Dialogue Concerning the Two Chief World Systems*. This work, the *Dialogue,* was put upon the *Index Librorum Prohibitorum*, List of Prohibited Books, which restricted the circulation

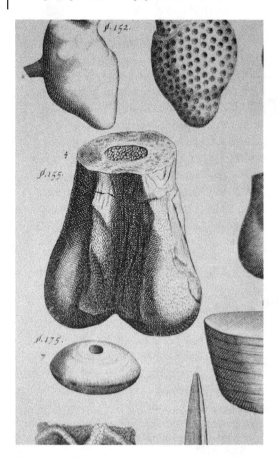

Figure 3.3 The first illustration of a dinosaur bone, from The Natural History of Oxford-shire by Robert Plot (1667). Although Plot considered this to be a gigantic thigh bone of unknown origin, he conceded others thought it to be a coincidental product of nature.

among Catholics. Much later this also contained *Zoonomia,* by Erasmus Darwin, Charles Darwin's grandfather. Perhaps surprisingly none of Charles Darwin's books ever appeared on the list.

What Plot was trying to do was explain observations within the context of a strong personal belief. It was for this reason that Plot was attempting to find plausible explanations for the fossils which he found extensively in Oxfordshire. He could not deny the existence of the things he saw, and as there was no accepted current thought on how they had got where they were, or what they were, it was possible to create a system of paradigms that fitted both the facts and the prejudices. *The Natural History of Oxfordshire* was a very influential book when it was originally published in 1677, and it was sufficiently popular to be reprinted in further editions. These later issues included a short biographical note by Edward Lhwyd, Keeper of the Ashmolean Museum, who had started as assistant to Plot, taking over from him as Keeper in 1690 until his death in 1709. There were also additional notes at the end of each chapter by the stepson of Plot, John Burman. Even though the title of the book implies a general guide to natural history, the content is more a learned treatise on the county as a whole, from the waterways and soils to the birds, plants and architecture. Littered with phrases in Latin and Greek, even though it was written in English, it was for those who had received a classical education. In the case of the Greek it is in Greek script, making it quite tricky for the modern reader. The work was well received, and his illustrations were taken from specimens in his own collection, but sadly none of his original specimens as illustrated have survived, which makes direct comparison with the engravings impossible.

Plot started from the premise that the fossils he was looking at, many of which are identifiable now using his illustrations, were not fossils in our sense at all. He also suggests that petrification from certain types of water can be a significant aspect of found objects. This process as he described it is quite common in the areas which have such chalk bedrock as is found in Oxfordshire. In the early chapters, he alludes to this process as occurring in many situations, including to shells. This precludes his more complete treatment of fossils later on, which follows directly from his discussion of the geology of the county and the various types of stones which can be found. Here the term 'stones' covers the various minerals and deposits that can be used for all manner of functions, from building to road surfaces. During his chapter *Of Stones,* which precedes the chapter covering fossils, Plot quotes the author Agricola (originally Georg Bauer) (1494–1555). This is particularly interesting because Plot makes reference to *De Re Metallica* (On the Nature of Metals), published in 1556 a year after the authors death rather than *De Natura Fossilium,* which was published many years earlier. It is *De Re Metallica* for which Agricola is most well known, so it is no surprise that the Keeper of the Ashmolean Museum would be familiar with it, we also know that since it was not translated into English until 1912, Plot must have read the book in Latin.

Plot's description of fossils comes into the chapter with the title of *Formed Stones*. He distinguishes formed stones from any other because they are naturally formed and seem to be, as he puts it, more for admiration than use. He does modify this by saying there is no way we could know what future us may be found for them. Not only were the descriptions produced of interest but also the drawings of the specimens were often of sufficient detail and clarity that we can identify the type of organism that they illustrated. A good example of this is his description of star-stones or *Asteriae* (Figure 3.4). This latter term, although as far as Plot was concerned it was a simple description of a formed stone, was fortuitous in

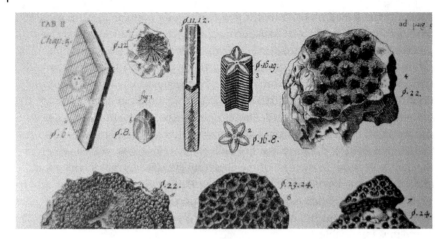

Figure 3.4 Centre top, Plot's Asteriae or starry stones, later recognised as Crinoidea (Echinodermata). *Source:* The Natural History of Oxford-shire (1667).

mirroring the Asteroidea in name and taxonomic position. It is now thought that the stones he illustrated were a species of sea lily the genus *Isocrinus* from the Lower Lias, which are members of the Crinoidea in the Phylum Echinodermata, the same phylum in which the Asteroidea are found. Although the naming is coincidental, it reflects the same observational naming that taxonomists later used because Plot did not accept that these were the remains of living things. Although he had his own opinions on this aspect of formed stones, he clearly acknowledged the opinion of Mr. Lister, as published in the *Philosophical Transactions* of the Royal Society that these very same stones were the remnants of petrified plants, although also by his admission, not of any species which had as yet been found either above or below the water mark of the sea.

Plot next goes on to describe similar structures in rubble-stone (Figure 3.5), which is now often referred to as coral rag. This is predominantly made up of limestone created from ancient coral reefs. The remnant coral becomes cemented together rather than remaining as separate units and becomes to all appearances a single block. Again, such is the precision of the illustrations that we can be fairly sure, taken in conjunction with the known geology of the area and what can still be found now is that the two predominant species he drew were *Isastrea explanate* and *Thamnastrea concinna,* two coral species. Among the illustrations by Plot is one of a mollusc that is now known to be of *Pseudomelania heddingtoniensis* (Figure 3.6). This was described as a species in 1880 by Blake and Huddleston and named in honour of the site at which Plot found his specimen, Heddington quarry in north Oxford. This spelling has changed over the years and is now universally Headington.

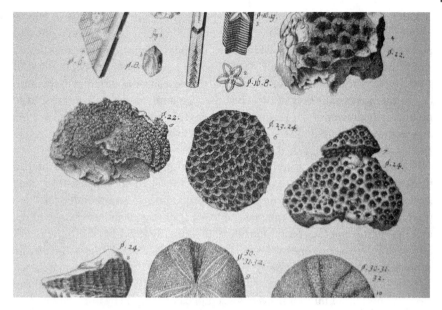

Figure 3.5 Rubble stone, coral concretions, that Plot describes as 'dug only for mending the High-ways'. *Source:* The Natural History of Oxford-shire (1667). Number 6 Thamnastrea concinna Number 7, Isastrea explanate.

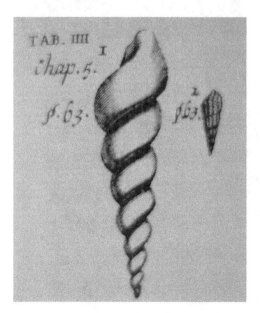

Figure 3.6 Pseudomelania heddingtoniensis, named after the original site of discovery, Headington in Oxfordshire.

Although the changes in place name spellings are understandable as so few written examples would have been available for reference until road signs and place names on maps became widespread, changes in spelling of more ordinary words are equally commonplace. Such was the random style of spelling that in 1746 a group of London booksellers were so perplexed by the variation in spellings of routine words that they approached Samuel Johnson with a commission to produce a dictionary of the English Language. This was finally published in 1755 and remained the pre-eminent dictionary for 173 years when the Oxford English Dictionary was first published.

Plot then goes on to structures found in the ground sent from the 'inferior heaven' as he puts it. This is a reference to structures that are supposedly formed in clouds and discharged during times of violent showers and thunder. Of these products of the atmosphere the ones he looks at in most detail are fossilised sea urchins, noting that they bear a considerable similarity to *Echinus* with the spines removed. Interestingly, one of the still common fossil echinoderms of the middle Jurassic from the Cotswolds is *Clypeus ploti* (Figure 3.7), it is being named much later in honour of Robert Plot, having been very clearly illustrated in his *Natural History*. This idea of inferior heaven is a complicated one and by implication there

Figure 3.7 *Clypeus ploti*, named in favour of Robert Plot after the original illustration in The Natural History of Oxford-shire (1667). *Source:* The Natural History of Oxford-shire (1667).

is also a superior heaven. Broadly speaking, spiritual principle is interior and part of superior heaven, while intellectual principle is external and therefore inferior. In theological circles, these definitions are highly variable and open to considerable interpretation, so precisely what Plot meant is open to question.

Although we may see it as self-evident that items such as sea urchin shells were organic in origin, it should be realised that there were many stones which had a striking resemblance to organic forms but which we would not begin to accept as fossilisations of the depicted species. It would only take a small step to apply the same principle of similarity without causality to more complicated fossilised material. Plot gives us two examples of this, one of a piece of flint that has a striking resemblance to an owl and another where he draws our attention to the similarity between a stone and a human foot. Admittedly this would have to be a foot in a sock, but a foot none the less.

What we find in the writings of Plot are descriptions of various shells, clearly being treated as if they are bivalves of various kinds, while not acknowledging that they are literally bivalves of organic origin. Interestingly, before Plot starts to describe his ideas on the origins of formed stones, he acknowledges the dissent that both Robert Hooke, spelling his name Hook, and John Ray expressed for any non-organic origin for these fossils. Both of these authors were writing only a little while before Plot. Robert Hooke gave a detailed account of his thoughts in *Micrographia* published in 1665. On page 111 Hooke says

> From all of which, and several other particulars which I observed, I cannot but think, that all these, and most other kinds of stony bodies which are found thus strongly figured, do owe their formation and figuration, not to any kind of plastick virtue, inherant in the earth, but to the shells of certain shel-fishes, which, either by some Deluge, Inundation, Earthquake, or some such other means, came to be thrown to that pace, and there to be filled with some kind of mudd or clay or petrifying water or some other substance, which in tract of time has been settled together and hardened in those shelly moulds into those shaped substances we now find them; ...

This, you will note from the end of the passage, is not the end of the sentence, which like many contemporary works contained single ideas in a single sentence whenever possible. In 1673, John Ray published *Observations Topographical* (1673), where he quotes large passages from Hooke but also gives additional reasons for considering fossils to be animal in origin and speculates further on how they appear where they do, far from the sea and high in the hills. So he says that if they were scattered by a rain caused flood, it would be expected that the shells would gravitate towards the sea, rather than move upwards. On the other hand, a flood caused by a deluge, such as a dam burst, would scatter the shells. Then he looks at the possibility of earthquakes raising the level of the land and concludes that generally it might be possible to raise the height of hills, but not to the heights of the Alps. On pages 126–127 he writes.

> If the mountains were not from the beginning either the world is a great deal older than is imagined or believed ... he then goes on ... or that in the creation the earth suffered far more concussions and mutations in its superticial part than afterwards.

That something was in operation, which was broadly not understood, was an astute observation but also, the shells so commonly found were not *similar* to sea shells in appearance, they *were* sea shells.

Notwithstanding these earlier, but essentially contemporary comments by both Hooke and Ray, Robert Plot devised explanations for most of the fossils, which were found in great number around Oxfordshire, that did not accept them as organic in origin. Plot obviously saw this as an important contradiction as he inset the entire paragraph made up of two questions which dealt with this as if to emphasise this particular point. This is quoted below.

> Whether the stones we find in the Forms of Shell-fish, be Lapides sui generis, naturally produced by some extraordinary plastic virtue, latent in the Earth or Quarries where they are found? Or, whether they rather owe their Form and figuration to the Shells of the Fishes they represent, brought to the places where they are now found by a Deluge, Earth-quake, or some other such means, and there being filled with Mud, Clay, and petrifying Juices, have in tract of time been turned into Stones, as we now find them, still retaining the same Shape in the whole, with the same Lineations, Sutures, Eminencies, Cavities, Orifices, Points, that they had whilst they were Shells?

He does write immediately afterward that he has not made a peremptory decision on this, but will debate the issues, although he states he is inclined to the opinion of Mr. Lister that they are *Lapides sui generis*. This Latin phrase broadly translates a 'stones unto themselves', meaning they are stones, not remnants of something else, certainly not formed from living creatures. Part of the reason for this referencing Lister as an authority was that he was recognised as the foremost authority in conchology of his day, we now recognise something more; that he was the originator of the subject. Consequently, his voice on the subject of fossil shells carried considerable weight. By assuming that fossils, as found, were a consequence of an unknown force, the question arose as to how they came in the forms in which appeared.

Plot considers this after detailing reasons why floods and deluges would not account for finding these shells at high altitude, or simply so far from the sea.

As to the purpose of formed stones, Plot has a comment on that. He suggests that just as little is known of the use of flowers such as Tulips or Anemones, they must have been created to beautify the world and consequently the same may well be true of formed stones.

Plot recognised that salts were the principal ingredient of stones of all sorts, with the type of salt giving the structure, solidity and duration of a stone's existence. Quoting one of his colleagues, Plot recounts the formation of various crystal structures in drying salt solutions analogous to ice formations on a window, or the snowflakes described by Hooke in *Micrographia*. The salt from which such patterns that can be found in formed stones were of some considerable interest to Plot. For some of the formed stones that were described by Plot, he suggests that urinous salts could account for the structures which are found. This was most specifically the patterns in many of what we recognise now as marine bivalve molluscs, and he also made a comparison with patterns illustrated in *Micrographica*. Plot also makes reference to stones that have a close similarity to fruit or fruit stones, these we can be sure are formed stones, probably alluvial, having all the appearance of being rounded and polished by the long-term action of water.

By taking the position that his formed stones did not generally have an organic origin, it was easy for Plot to describe one of his finds, which he describes as commonly found, as *Hippocephaloides* since it resembles the skull of a horse, albeit a very small one. This name refers specifically to the appearance of a type of stone and all those like it and does not imply that he was suggesting any equine relationship. From his illustration it is most likely that this is in fact a cast of the shell of the marine mollusc *Trigonia,* an extinct genus of saltwater clams (Figure 3.8). This process of naming stones based on their recognisable shapes extended to rocks of a random kind, like his single stone which he calls *Auriculares,* shaped like a human ear. The human ear is only one among many of the stones Plot collected to which he gave names directly related to human body parts, both internal and external.

There was one formed stone that confounded Plot, until after long consideration he assessed it as a petrified bone. Given that it was bone, it was apparent to Plot that this was part of a femur and not an accidental similarity to a bone of organic origin. Since he accepted that it came from an animal the next step was to spend some time on investigating what species it originated from. It was at this point that it became necessary for Plot to fit it into a zoology of known species; it had to be a living form as there was still no concept of extinction acceptable to his theology.

Starting from the premise that petrification did not alter the dimensions of the bone, he immediately determined that it was neither cow nor horse, having to be an animal much larger than either of these. His next consideration was that it might be an elephant, possibly brought to Great Britain by the Romans. This he thought unlikely as no Roman author had recorded an elephant in Britain. Also, it would be assumed that if it was an elephant bone, then the more robust teeth

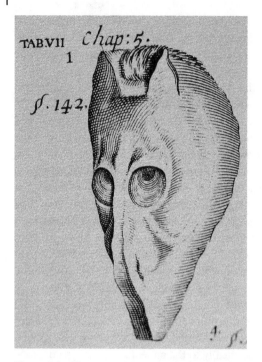

Figure 3.8 The stone referred to by Plot as Hippocephaloides, resembling the skull of a very small horse, and now thought to be a marine mollusc of the genus Trigonia. 'I have taken the boldness to fit them with a name, and in imitation of other Authors (in the like case) shall call them Hippocephaloides'.

and tusks should be found at the same site. The clinching evidence for Plot that this was not an elephant bone came when, as he was writing his *Natural History of Oxford* in 1676, a living elephant was brought to Oxford for show. He compared the thigh bones between the elephant and his specimen, recognising that it was not only a different shape, but that the elephant thigh was bigger than the fossil, even though the elephant was not fully grown. Although he paid some attention to the size difference, it must primarily have been the shape which convinced him of the difference as his fossil femur may also have been from a juvenile individual.

At this point the otherwise curious and investigative Plot moves back to secondary sources and his theological background. He starts to build a case for this bone to be human in origin by quoting biblical references, such as Goliath being 9 ft 9 in. in height, which is about 2.97 m. He also describes writings from the classical world where giants of many types can be found (see Chapter 2). At the time of Claudius, a giant, who by coincidence was the same height as Goliath, by the name of Gabbara was brought out of Arabia. This would put the fossil and Gabbara at about the same size, so he surmises that Claudius could have brought Gabbara to Britain where, to complete the story, he would then have had to die, leaving the petrified thigh bone for Plot to discover. He lists further examples such as Eleazor who was 7 cubits in height, that is about 3.2 m, and writings of Goropius Becanus, physician to Lady Mary, sister to Emperor Charles V, who claims that living in his vicinity was a man a full 5 m tall. From his considerable trawl through ancient writings and hearsay reports, he concludes that the petrified bone is a genuine human thigh bone, but from a giant. The circumstances and geography of the discovery of this bone as well as the illustration of it would indicate that it was the femur of a broadly bipedal therapod dinosaur, probably a *Megalosaurus*. These are thought to have been between 1000 and 3000 kg, which makes it a large animal, although much smaller than an African Elephant that can grow to 6000 kg.

The way in which Robert Plot tried to reconcile what he was seeing in the fossil record and his world view, powerfully influenced by his religion, is of some interest. This was not unusual for the time, as it comes through in many early palaeontological stories. Even though the observational part of the science, the results of which cannot be gainsaid, was clear, deference to a religious upbringing tempered the explanations. This mixture of explanations, where the observed facts were tailored to match preconceptions broadly fits the system of thought sometimes referred to as Neoplatonism. This is a modern description, and the Neoplatonists of the time would refer to themselves simply as Platonists.

This group of philosophers, the Platonists, saw interlinked affinities between all things, so that there could be a pervasive moulding force or 'plastic virtue' governing the growth of living organisms. In a similar way, Plot considered plastic virtue as a form of crystallisation, he thought this pervaded all things and allowed rocks to form the same structures as living things. This concept of plastic virtue being all encompassing is as far as Plot and Neoplatonists would agree. This was especially so as the original Neoplatonists had been the last bastion of anti-Christian philosophy in the ancient world.

It is worth a note that although Francis Bacon had published *Novum Organum Scientiarum,* New Instrument of Science, in 1620 and while his tenets of reductive and inductive reasoning were being enthusiastically embraced by Isaac Newton and other contemporary scientists, it was not the central idea used by Plot in his investigations. Bacon suggests a sceptical mind to assess observations and by implication hearsay reports, while Plot takes the ancient texts at face value, that is, literally true. This automatically renders any conclusions which Plot comes to, based on such shaky foundations, questionable. Even if it should turn out that the ideas put forward by Plot were correct with the edition of later evidence, by virtue of his line of reasoning, it would have been reasonable for his contemporaries to doubt his conclusions. His was a more syllogistic method, which can by its very nature create contradictions and inaccuracies.

While debate in the seventeenth century regarding the origin and significance of fossilised material was starting to move towards an overall acceptance of an organic origin, this was based on a developing body of scientific knowledge. Nonetheless, these ideas were couched in terms which protected the writer from any suggestions of heresy. There was continuing antipathy by the Church towards such powerful ideas that were intrinsic to such simple enquiries as to what fossils represent and how old they were. This was a continuing problem for developing scientific thought. It was not intrinsically a problem of fossils, it was the associated implications, such as extinctions and biological change. These were the ideas that were so difficult to fit into Biblical dogma. This situation was carried on into the eighteenth century and beyond, sometimes with a considerable influence on the dissemination of ideas. One such example of this is the work of Benoit de Maillet (Figure 3.9), a devout Catholic who lived from 1656 to 1738. His main

BENOIT DE MAILLET *Gentilhomme Lorrain, Consul General du Roi en Egypte et en Toscane, depuis Victeur general des Echelles du Levant et de Barbarie ; et nomme par Sa Majeste en qualite de son Enyoye vers le Roi d'Ethiopie ; Auteur des Memoires sur l'Egypte, et sur l'Ethiopie .*

Figure 3.9 Benoit de Maillet (1656–1738) depicted in about 1735 in armour as an indication of status as a nobleman, rather than as an indication of profession. *Source:* Étienne Jeaurat, https://en.wikipedia.org/wiki/Beno%C3%AEt_de_Maillet#/media/File: Beno%C3%AEt_de_Maillet.jpg, Public Domain.

work, the publication for which he is best known is titled *Telliamed,* which is his name simply reversed. Although the content of *Telliamed* was well known, it had a rocky history because of ecclesiastical interference in its production. *Telliamed* was published posthumously in 1748, translated into English in 1750 (de Maillett 1750), and demonstrates the clarity of thought and difficult conclusions which de Maillett came to. It was delayed in publication by being passed to Abbot Jean Baptiste de Mascrier for editing. He had worked with de Maillet on a previous work, but the affect he had on *Telliamed* was so damaging that when it was published it did not correspond very closely to the original ideas that de Maillett had wanted to put forward. It took several centuries before the convoluted problems associated with rewriting by de Mascrier were unravelled. The attempts to get back to the original ideas culminated in a translation of 1968 which took the radical step of taking for a starting text the early manuscripts of various sorts which were circulated between 1722 and 1723.

The premise of *Telliamed* is a conversation between an Indian philosopher, who is the eponymous Telliamed, and a French Diplomat. It is thought that de Maillet hoped that using this, frankly transparent ruse, he could escape the repercussions which would be associated with publication of a heretical tract of this sort. The major problem with the manuscript as it was published in 1748 was that de Maillet had clearly appreciated that the Earth had not been created in an instant. He came to this conclusion by observation of geological features that are visible for all to see and interpret. He also managed to take one step further and believed that land animals had originated from salt water ancestors. These two ideas were in such direct contradiction of Catholic teachings, and a literal belief in the bible that before publication Abbot Jean de Mascrier decided to heavily edit it. He tried to reconcile the whole book with the teachings of the Catholic Church, which of course was going to be virtually impossible without completely changing the ethos of the work. The only way it could be made to satisfy the religious convictions of Mascrier was by emasculating and changing the text almost beyond recognition. This was not a philosophical document open to debate, there were too many carefully worked arguments based on precise observations, it was an idea which was liable to undermine the very foundations of literal belief in biblical Christianity.

De Maillet was a proponent of ultraneptunism, a modification of neptunism, which suggested that rocks were laid down by crystallisation of salts out of the ocean. Ultraneptunism as described by de Maillet was based around a process of sedimentation rather than just crystallisation, with weathering playing a very small part. Although the sedimentation process is significant, he did not comment on the possibility of other methods of creating different rock types. Being a believer in modified neptunism, de Maillet understood that the organic origin of fossils could easily be fitted into his concept of rock formation. This had all sorts of implications, but most importantly, that fossils did not arise as spontaneous images within sedimentary deposits, but were formed in an as yet poorly understood way from the original animals and plants. This small but clear idea is of great significance because starting from the point of current knowledge as it was known at the time, it took considerable insight to see that these very definitely solid rocks came from loose material, suspended in the oceans. The creation of a natural cement was hard to conceive of, so although the method would remain unknown and consequently could not be speculated about, the outcome was there to be seen in every limestone quarry. With the understanding that fossils were formed by deposition of sediment, it became self-evident to him that the planet was originally covered in water. This was not an entirely original idea as it had been considered by Rene Descartes a century earlier. What de Maillet did was take the idea of a watery planet and follow it through to a conclusion. His logical argument was that given the planet was originally under water, he calculated the rate of sea level fall was about 7.5 cm/century. This value was arrived at by looking at historical ports that had long before been by the sea but were now some distance from the coast. Based on this calculation de Maillet also suggested that the age of the earth was about 2.4 billion years. This very long time scale caused considerable problems within the Church and also introduced the idea of geological processes being very slow indeed. The difference between de Maillet and Descartes stems from Descartes being in thrall to the power of belief engendered by the Church. Even so Descartes had problems with his writings being accepted in their entirety. This antagonism with the Church culminated in 1663 when many of his works were indexed by the Catholic Church, this involves putting them on the published list *Index Librorum Prohibitorum*, the index of forbidden books. As draconian as its title implies, indexing was only officially dropped in 1966, although the last edition was published many years before. It should be remembered that Descartes was in good company as around the same time various bans were imposed on Pascal, Calvin, Bacon and Spinoza.

Descartes considered that stars, as suns, could die by accumulation of a crust. This he thought was probably associated with surface sun spots as they appear darker, as though they were forming a crust. These stars could then wander the skies until caught by another star where they went into orbit to become a planet. The relevance here is that Descartes' hypothesis was used as an idea in *The Sacred Theory of Earth* by Thomas Burnet, published in 1691 (Burnet 1691). Written with the confidence and certainty of the seventeenth century cleric and with considerable embellishments, *The Sacred Theory of Earth* tried to explain all such observations in biblical terms. In this book, 'The Water of the Planet' was held under the crust, which supported the garden of Adam and Eve. The advent of the Deluge was the collapse of the crust, the water replacing the land. It was only with the account of Newtonian cosmology that this complicated and inexplicable explanation was replaced by something both more tangible and understandable. The ideas concomitant with developing cosmology also implied that the earth really did have to be older than biblical time scales as stated by the Church. This new mechanistic attitude to cosmology could be just about fitted into the framework the Church insisted on, by saying that God created a system and the system created the planets. It was still heretical in the final outcome, but accruing knowledge was making the new sciences of astronomy and geology difficult for even the entrenched Church to ignore.

The *Telliamed* was also heavily influenced by the ideas of Descartes, although access to the works would have been limited by some of them being *Indexed*, either in their original Latin or when they had been translated into French. The reductionist scientific method of trying to describe the origins of the planet earth, its life and fossils as a complete story, rather than a discontinuous series of piecemeal events was gaining ground and would not easily be reversed.

With the increasing evidence that fossils were the remains of living things, a further complication arose that previously had not been considered. This was that prior to this point the concept of extinction was unknown, but by implication, it now had to be embraced. If fossils represented living species that no longer existed, then they must have become extinct and this ran contrary to biblical teachings of perfection in living things rendering them immutable and permanent. Of course, the other problem was in the word 'immutable' because there seemed to be relationships and changes happening through time reflected in the living world as well as the fossil record. As taxonomy had developed from Aristotle onward, it became increasingly obvious that species were related and although that did not automatically imply change, it did raise the question of in what way the species were related. Consequently, did these related species have a common origin from which all other similar species arose by change? Of course, extinction had been going on during this formative period of science, but it was not recognised as such and its significance certainly was not readily appreciated. An illuminating example of this can be found in the well-known case of the Dodo (*Raphus cuculllatus*), a relative of pigeons from Mauritius. The Dodo was only known to Europeans for about 90 years before it became extinct in 1681, and at the time it disappeared with virtually no comment. Extinctions such as this can be seen on many remote islands as the pre-human fossil record shows that they often had unusual and exotic fauna, often with no mammals present. This is particularly noticeable in the Mascarene islands, of which Mauritius is part. The true nature of Dodos is unknown as descriptions vary widely and the only complete specimen belonged to the Ashmolean Museum. So little was thought of such things that it was decided in 1755 to dispose of it. Before incinerating the carcass someone removed the right foot and the head, these now being the last remaindered memento mori of the species. This attitude was very much kept going by a belief that just because you could not find it on one island, it will be present on another, extinction was still poorly recognised. There was another aspect that is sometimes overlooked, this was the perceived pre-eminent position of man implied that the world and all of its life was to be dealt with as he saw fit, care was not include in this model. It is now thought that the Dodo would probably have remained in obscurity, just another human extinction like the Tongan Giant Skink (ex. 1827), or Florida Black Wolf (ex. 1917), were it not for Lewis Carroll making it a character in *Alice's Adventures in Wonderland* (Carroll 1865). Hunting to extinction, or human intervention to extinction, seemed to be perceived as quite different to the idea that Creation had in some way made a mistake and species had become extinct without any human action.

It became apparent to many observers of the rocks and fossils that could be found quite widely and that there were distinct implications both for the age of the earth and how it had been structured in the past. Also, and most significantly as Robert Hooke noticed, some of these easily found fossils had no living counterparts, although he did note that ammonites had a passing resemblance to nautilus. This was an observation which in itself created potential problems for biblical stories as the question arose as to the relationship between the current and the extinct. These works by Hooke and his contemporaries were multidisciplinary publications in areas that were yet to be recognised as scientific disciplines in their own right, geology and palaeontology. They were not aimed at undermining Genesis in an increasingly mechanistic and secular world, but that is what they helped to do as a by-product of scientific enquiry. They were searching after truth and for the investigators little or no consideration was given to what were to be the unexpected consequences of their work. Even so, it was still going to take a long time before a universal agreement was achieved as to the nature of fossils and their significance to the age of the earth.

Such was the controversy that in 1695 *A Natural History of the Earth and Terrestrial Bodies, Especially Minerals,* was published by John Woodward, Professor of Physick, Gresham College (Figure 3.10). This was significant enough for it to be produced in several editions, the second in 1702 and the third in 1723. The main theme of the three parts to the book is an apologia for the biblical story of the flood. The bulk of the text is made up of explanations for biblical events. Woodward asserts, for example, that 'the whole world was taken to pieces and dissolved at the Deluge'. Shells and similar items are deposited with other material in the water, depending on their gravity, the heaviest being deposited first. This was an idea that was mentioned early on in the book and like many other ideas it was developed later on. One of his first statements is that observations are the only sure grounds upon which to build a lasting philosophy. While all of his practical observations were made in England his work was uncritical of the primary supposition upon which he based his ideas, the Deluge of Noah, and for which the only

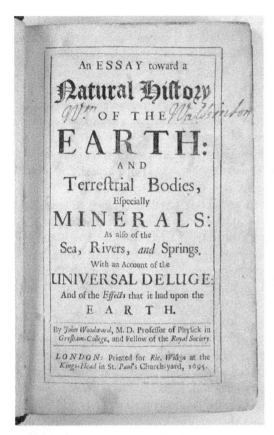

Figure 3.10 Title page of A Natural History of the Earth and Terrestrial Bodies, Especially Minerals, by John Woodward, published 1695. A prime example of an argument based on an unsubstantiated primary assumption.

reference is a third person narrative translation. At the start of the work, he hopes to dispel the idea that fossils in the form of sea shells are simply mimicking rocks, acknowledging that finding sea shells, in fossilised form embedded in rocks, implies they are rocks themselves. In support of this idea, he described belemnites. As these were of completely unknown origin at the time, it was easy to imagine they were formed stones, not only that, but they could easily have been formed considerable distances from where they were found. As a subject for investigation belemnites were particularly fortuitous as the exact nature of these common fossil remains of the internal skeleton of cephalopods, would remain hidden for many years. Woodward was trying to describe fossils as being organic in origin, although there were stumbling blocks, but here was a significant move towards a clearer understanding of the structure of the Earth.

Following on from other observers, he suggests that shells were the moulds for fossils formed in the cavities of the shells, this we know now to be the case in some circumstances. Part of his argument for this was that the dissolution of real sea shells should follow the exact same process as the dissolution of fossil shells. As we have noticed before, one of the traditional stumbling blocks of the story of the flood was that some fossils do not seem to have any current comparable species, and yet extinction should not happen in a perfect creation of organisms perfectly formed for their life. Notwithstanding this problem, Woodward recognises something most important that these fossil shells have all the necessary characters of being sea shells and show a relationship by their shape and structure as near to some now extant species as those extant species do to one another. As a consequence of this observation, he surmises that it was likely that there must have once been and still are, such shell fish in the sea, but at such depths that they remain unknown, since, as he puts it, pearl fishers and divers deal with shellfish that lie perpetually in the deep. So by making a leap of thought he assumes that those we can match, that is, fossils to living forms, and those which we cannot, are all remains of the Universal Deluge. Woodward goes on to claim, with a confidence in his authority that belies the lack of evidence, that those fossil species which cannot be matched with extant species must be too deep to find and are still living *in the huge bossom of the ocean: and that there is not any one intire species of shellfish, formerly in being now perished or lost.* This speculation regarding the existence of otherwise unknown species in the seas is hardly surprising since it was not until the nineteenth century that pumped air systems for sea floor workers were available. Even then, these were not generally used for biological investigations. It was not until 1943 when Gagnan and Cousteau invented the scuba system that investigation of the near coast sea floor could begin. Until then it was only via dredging that sea floor species could be brought to the surface and the depths of the ocean remained a completely unknown area.

By asserting that fossil material really did represent unknown species, the question remained as to how these organisms, embedded as they were in stone, could be accounted for. How exactly did they get there? Embedding, rather than infiltrating, was essentially correct in that Woodward described a process of deposition of sediment as the deluge receded. By his argument, the point at which the fossils became embedded depended upon the specific gravity of the suspended items. As a consequence of this, the lighter material will appear higher up the strata and over the 4000 years since the deluge these very same top deposits would be worn away.

Woodward suggested that changes to the surface of the earth were both slight and almost imperceptible. He was using this argument specifically to demonstrate that the bounds of the sea and land are fixed and permanent, which meant that the globe was exactly as it was immediately after the Deluge. This line of reasoning is interesting as it starts from the assumption of factual accuracy for his idea that change is small and therefore imperceptible. It follows from that idea that after the flood had receded the coast line did not change in any significant way. He does make a note of finding fossilised fish in the UK that are now only known from the coast of Peru, although he does not say what these species are. No particular comment is made regarding the significance of this, or the way that this strange situation could have occurred. He makes similar comments on various species of land animals, such as elephants, known as fossils in the UK and as animals elsewhere. The arguments put forward sound cogent, but we are never given any clear idea as to how some of his suggestions could come to fruition. An example of this is his statement that amber was formed in the Deluge, embedding insects at the same time. This is almost put forward as a statement defying contradiction.

Woodward takes a lead in dating the Flood from Bishop Ussher of Armagh, not by setting a date for the Deluge, but by setting it to a season. He asserts that with the large number of buds about to open found in peat bogs and marshes, the flood must have taken place in late spring. Accordingly all of the fossils and associated material would be lifted by the flood, which then deposited the material by selective precipitation according to the laws of gravity. These had been worked out and published by Isaac Newton in *Principia* only a few years earlier in 1687, making use of them giving a hint of modern thinking and additional plausibility. Woodford assumes the heaviest shells will be deposited first, with an acknowledgement of the density of shells, even though this simple reduction of mixed material to a single product, does not tell the entire story. Sedimentation of solids in moving flows can be quite complicated, being influenced by density and shape as well as size, and in the case of material about to become fossilised there would also presumably be organic matter from the living organism present as well.

As with all contemporary accounts in which fossils played a part, the subject of large-scale geological formations were also discussed, since these seemed inextricably linked to the position in which fossils finally came to rest. In this case, Woodward put forward the hypothesis that hard stones create mountains, with rocks rearing up and supporting each other, whereas gravels and clays are more 'champagne', this being an old English term for open and level country. His point of reference for this speculation remains biblical in origin. He makes a reference to an apparent anomaly in the assertion that mountains cannot be raised by earthquakes. This anomaly was a mountain in Lake Lucerne, not far from Pozznola, called Monte de Cinere, now known as Monte Nuovo. Woodward described it as a confused heap of stones. We now recognise it as a cratered volcanic cone and represents one of the first examples of an eruption of this type which was both witnessed and described by several people. The eruption took place at the end of September and beginning of October 1538, preceded by earthquakes, it has remained dormant ever since.

Differentiating between heat and fire Woodward developed an explanation of rain as recycled oceans. Using the same train of thought he surmised that earthquakes could be caused by a form of subterranean natural gunpowder. In broad terms *A Natural History of the Earth and Terrestrial Bodies* was one of the last attempts to completely reconcile the literal interpretation of the bible with observed facts (Woodward 1695). This, of course resulted in vast holes in his reasoning, where it was simply not possible to reconcile the observations with ancient stories without lapsing into the same illogical jumps that had been used by ancient scholars. The difference being that the ancients could ignore the laws of physics as they were simply unknown. Interestingly, there seems to have been a resurgence in these ideas with the rise of creationism and advocates of the 'young earth' idea. Still, I suppose science is too difficult for some people.

When Nicolas Steno was writing it was contemporaneous with Woodward, but otherwise quite different. Steno was a convert to Catholicism, eventually rising to the position of Bishop. This close affinity with the Church was quite likely a source of considerable friction, maintaining a clear association with his religion while dealing with his scientific observations. Although employed in the Church, Steno was scientifically very curious. It was this curiosity that was fired when he was looking in some detail at a sharks head. It struck him as significant that the shark's teeth were identical to the inclusions in stone at that time known as *glossopetrae* – tongue stones (Figure 3.11). He studied the differences in considerable detail and finally concluded that the chemistry of fossils could be altered without affecting the morphology of the original structure (Steno 1667).

It also lead him to wonder how one solid can arrive inside another. This particular enquiry extended beyond fossils to other geological features, but also other apparently similar situations, such as vessels within the body. These investigations and other studies in geology were published in 1669 as *De Solido Intra Solidum Naturaliter Contento Dissertationis Prodromus* (Steno 1669). This was the last and in many ways most significant work that Steno produced and although not the first to assume fossils to be organic remains of once living animals, he was one of the most significant clergy of the day to start the non-secular move towards recognising fossils as animal and plant remains. *Prodromus* was written in Latin, but in 1916 was produced in a very good English version. Although Steno concluded that fossils were organic, he still fought to fit the overall picture into the biblical story of creation. He tried to do this by claiming there must have been two periods of deposition, the first immediately after creation of the world and the other one as a direct result of the Deluge. The change in the strata, which should have been uniform and worldwide when they were laid down, could be explained according to Steno, by erosion and distortion of the strata associated with movement of the earth's surface.

While up to the time of Steno, the ecclesiastical natural historians had been developing a balancing act between observation and dogma, there was a developing culture that was going to alter the way science was carried out. It was also going to continue to annoy the entrenched and inflexible. This change was the

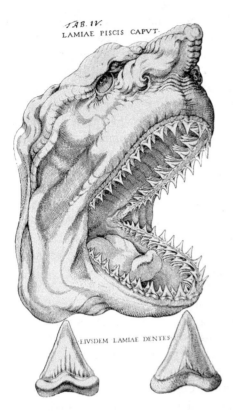

Figure 3.11 The association between glossopterae and shark teeth as shown by Steno in Canis Carchariae disectun Caput (1667). This is a reproduction of an illustration which originally appeared in Metallotheca (1719) by Michele Mercati (1514–1593). The disparity in dates reflects the first publication of a manuscript that was well known long before publication. It is worth noting that the image is an anthropomorphic shark, from a proper nose to a mammalian tongue. More of a human caricature with sharks teeth.

development of highly educated scientist who started by making observations and then tried to interpret them without having to make leaps in the dark based on preconceived ideas. For many it involved correct observation and measurement before anything else was considered. This started in the seventeenth century when science and engineering started to become less the activity of the dilettante and more the activity of men of vision. The new breed of enquirers who had the energy to state what they had found regardless of the consequences.

One of the early scientific observers in this new and developing scientific revolution was Robert Hooke. This was convincingly demonstrated with the publication of *Micrographia* in 1665 (Hooke 1665), but some of his later works are of equal significance in changing attitudes to the origin and importance of fossils.

Although presented at the Royal Society, the complete text only became available in 1705, Hooke having died in 1703. When it was finally published in complete form, it was as *The Posthumous Works of Robert Hooke*. One of the particularly interesting aspects of this was that many of his amanoid fossil specimens are still known and give testimony to the accuracy of his illustrations by being clearly recognisable. That these were clearly animal in origin seemed clear to Hooke, even though there were species present in the fossil record not known from living specimens. His were the first accurate descriptions of ammonites that were based on the principle that they were animal in origin and not a result of 'plastick virtue'. Interestingly Hooke describes 'spars, diamond, crystals and rubies' to be the 'ABC of Natures working' as they are easily explained mechanically. In comparing ammonites in general to *Nautilus,* Hooke made note of the differences between them, this observation formed part of his reasons for suggesting that fossils were of considerable importance. Another reason in his list of why fossils are organic, was that in former times there were species not found now and there are other species found now that were not found then.

In contrast to the method of Hooke, of observing and trying to work out the best explanation for what he saw, John Ray started his investigation of fossilised material in the same way, but later in life became overly influenced by the unsupportable doctrine of literal interpretation of the bible. This forced his hand into making observations, and reporting other people's observations, but then trying to make them fit a predetermined model, even when simpler explanations would work much better. In 1692, John Ray published *Concerning the dissolution of the World* (1692), in which he entertains a teleological notion of fossils by saying -

> that Nature doth sometimes ludere, and delineate Figures, for no other end, but for the ornament of some stones, and to entertain and gratifie [sic] our curiosity or exercise our wits.

His arguments as to the nature of fossils range quite widely in this book, including the idea that the deluge as described was too short to produce the huge alpine beds of shells. Interestingly, Ray makes a note of an ammonoid *Cornua Ammonis,* now an obsolete term, which he describes as 2 ft in diameter, but he claims it had never been related to any shells or fishes. He generally argued that unknown species were still alive and active elsewhere in the world. This line of thought was turned upside down when in 1693, John Ray published *Three Physico-theological Discourses,* dedicated to John, Archbishop of Canterbury, Primate of All England and Metropolitan (Ray 1693). This was reprinted many times by different publishers.

The *Discourses* dealt with many aspects of the world and its geology, which includes the perceived problems of contemporary thought when dealing with the nature and importance of fossils. It also tackled some biological phenomena that were currently unexplained at the time. In Discourse I, Chapter 4, there is an illuminating insight into the current thinking regarding perpetuation of species as unchanging units. He notes that there are two different classes of thought. The first is that every original animal contained the seed of every following generation, effectively giving a fixed number of generation for a species. The second is the idea that God gave every animal species the power of generation. Ray suggested that the first was correct since both Hooke (spelt Hook) and Leeuwenhoek (spelt by Ray as Lewenhoek) had seen tiny creatures invisible to the naked eye using microscopes.

The most interesting part of the *Discourses* for palaeontologists comes in Discourse II *Of the general Deluge in the Days of Noah, its Causes and Effects.* In this section, over a considerable length Ray outlines the ideas and consequences of various interpretations of fossils. It has to be realised that without a fossil record the literal interpretation of the bible would not cause any significant problems, or more precisely, interpretation of the geology of the world would not cause any significant problem for the literal interpretation of the bible. Fossils, it would seem, were a problem that could not be ignored. However, the debate was raging as to whether these formed stones, which were scattered far and wide, were original products of nature or imitations of shells and fishes. In this, he frequently makes use of the idea of 'plastick virtue', as disregarded by Robert Hooke.

Ray discourses at length on oyster shells and their being found in considerable beds, often very great distances from the sea. Ray thinks it unlikely that nature would create shells so obviously oyster-like with no intent of their being an animal involved. This was a conundrum for Ray that he finally came back to and resolved in a surprising way. He had the same problem with teeth and bones of fishes, having even seen fish scales preserved within stone. As he put it in *Discourse* II,

> 'Nature never made teeth without a jaw, nor shells without an animal inhabitant, nor single bones, no not in their own proper Element, much less in a strange one'.

It is interesting that he questions why only shells and fish teeth are found as stones, but no bones of land animals, or fruit, nuts or seeds. Ray was probably unaware of the ancient finds of the Mediterranean, but would have been aware of the findings of Robert Plot. Plot had misinterpreted his findings as not animal in origin, but as representing part of human anatomy. It was, therefore, reasonable that Ray would assume no land animals could be found within the rocks and stones where the beds of shells were found. This was another assumption that was going to be significant in the final conclusion that Ray came to.

One of the increasing problems that had to be addressed at this time in this argument was brought about by the increasing range of voyages of discovery and commerce. This was that the inquisitive travellers were discovering fossils in different places which did not correspond to any known species. Ray quotes one such discovery by Sir John Narborough. He stopped at Port St Julian in the winter of 1670 where he described beds full of oyster shells which could not have come from either sea or flood because there is no such shell fish in those seas or shores. This discovery was described in *An account of several late voyages and discoveries to the South and North towards the Straits of Magellan, the South Seas and the vast tracts of land beyond* (1694) by Sir John Narborough. Ray counters this with a simple statement that there might be such species, Sir John had just not seen them, or a tempest had brought in the shells. This line of argument is difficult to gainsay, and in fact this is the type of reasoning which became the basis of constructing the null hypothesis in later years.

The other aspect of finding shells in high mountains, which he finds unaccountable, was that there was an implication that some species would have been lost. Extinction was a real problem to the concept of perfect creation. As Ray puts it,

> 'It would hence follow, that many species of shell-fish are lost out of the world which Philosophers hitherto have been unwilling to admit, esteeming the destruction of any one species a dismembering of the Universe, and rendering it imperfect'.

Ray accounts for the unknown species by repeating the argument that unknown does not equal none existent. Also, the very large shell fish of the tropics are at least as big as any petrified species found, with the exception of the one he refers to as *Cornua Ammonis,* which he suspected *to have never been, nor had any relation to any shells or fishes.* He uses an interesting analogy for the unknown species by referring to lost species such as bears and wolves, which used to be found in Britain, but although no longer present, they were still common enough in Europe. These were relatively recent extinctions, officially the last wolf in Britain having been killed by Sir Ewen Cameron in 1680 in Perthshire while the beaver (now reintroduced), was hunted to extinction in England in the twelfth century and in Scotland in the sixteenth Century.

High mountain shell beds could, according to Ray, have been because originally, post Deluge, not all land was uncovered at once, so the remaining submerged areas could support a growing population of shellfish until this submerged land became raised above sea level. These arguments all pointed towards Ray's original belief that fossils and fossil beds had an organic origin, it is a surprise, therefore, to find that at the end of this discourse Ray reproduces a letter from Edward Lhwyd to John Ray. This, as far as we can tell, was an open letter, rather than a personal one as it appeared first in a publication by Edward Lhwyd, *Lithophylacii Britannici Ichnographia* (1699), which was a catalogue of 1766 fossils and mineral specimens, at the back of which there was a collection of six letters directed to various friends, the sixth of which was addressed to John Ray. This was a common technique in natural history writing, probably the most well known of which is *The Natural History of Selborne* by Gilbert White, first published in 1789 (White 1789).

The arguments put forward by Lhwyd are accepted by Ray such that although Ray does seem keen on the idea of a biological origin for fossilised material he shifts his position towards a non-animal origin. The letter that seems to convince Ray of this is spread out over 28 pages in the *Discourse* answering all of the queries put forward to suggest animal origin. Ray does leave the question open, suggesting that if anyone should know a better explanation they should make it known, almost as though the acceptance is a reluctant one. It certainly does not give the impression of being a simple change of sides in the ongoing argument as to the origin and therefore significance of fossils. He freely admits that raised oyster beds do cause some concern, he admits that they may have originated in the same way. The broad idea that Lhwyd had, was that mists and vapours from the sea were impregnated with spermatick of marine animals. These were then raised up out of the sea and carried long distances before descending to land where they penetrated deeply into the ground. Once in the ground they would germinate to make complete, or part complete, organisms in the stone. There they would stay until discovered by being washed out of cliff faces or in quarries. This was also used as the reason that fossils of species not known from Britain could be found, sometimes quite easily. There is no doubt that part of the weight of the argument for Ray came from Lhwyd being from the Ashmolean Museum, originally as assistant to Robert Plot and then a Keeper himself when Plot relinquished the post.

Perhaps the last attempt to discredit the hypothesis, by now regarded as a theory, that fossils did represent animal from past ages and that the age of the earth did not correspond with the biblical age as described by Archbishop Ussher, came from Philip Henry Gosse. While previous writers who had been trying to deny the accumulated evidence of geology, such as Plot and Lhwyd, had approached it piecemeal, Gosse took a broader brush to the question. By the time of publication of *Omphalos: An Attempt to Untie the Geological Knot* in 1857, the belief in the nature of fossils as organic was well established. *Omphalos* was the book that laid out the arguments for a young earth, taking its name from the central argument, according to Gosse Adam had a navel, although having been created new he did not need one, it was there to give the appearance of a history (1857). Published 2 years before the appearance of Darwin's *Origin*, *Omphalos* contains several mentions of Darwin, but never in a derogatory way, Gosse was an accomplished naturalist in his own right and was aware of the work of other scientists of the age.

Broadly, Gosse argues that fossils are not evidence of past times being different to the current age. We would now say he denied the evidence of fossils for evolution. Fossils, for Gosse, were the result of an act of creation to make the earth look older than it is. This was how the example of Adam came to be used and reflected in the title. This circularity of argument that all objects were newly minted with the appearance of a past extended even to the original trees having fallen leaves that were never on the tree. This line of reasoning extended to fossils and geological strata, they were created to give the appearance of an ancient origin. Needless to say the concept is a logical progression that is difficult to argue with unless the formative assumption is tackled; how do we know Adam had a navel, how did we even know Adam existed? The problem with the arguments of the sort Gosse was making is that the secondary product of the idea, that everything is put in place to make it look old, is almost impossible to dismiss simply because any counter argument can be discounted with the repetition of 'God made it look like that'. These flaws in the argument, invoking what is little more than magic, seem to have been recognised at the time as the book did not sell well.

By the time *Omphalos* was published the first clear statement for the age of the earth reflected by the fossil record had already appeared in print. In 1844, *Vestiges of Natural History of Creation* appeared, originally as an anonymous publication, but later known to be written by Robert Chambers (Figure 3.12). It was not until the 12th edition, produced in 1884, coincidentally 12 years after Chambers had died, that authorship was acknowledged. Although anonymous authorship was common in the nineteenth century, this particular example of anonymity was slightly different. Authorship was kept deliberately as *Vestiges* was regarded with great disdain by the clergy of all denominations, but more especially as a Scot in St Andrews he considered it as likely that the evangelical theologians of the Scottish church, had they known who he was, could have severely damaged or even destroyed his thriving family business.

The business, which Robert and his brother William had was Chambers, publishers, still extant as Chambers-Harrap, Publishers, producing a range of dictionaries. Even given the dislike of the content by the theologians of the nineteenth century, the book sold well and notwithstanding the contemporary criticisms of it, the book had considerable influence. The text starts with a large view of the probable origins of the solar system, but then goes onto the concept of original spontaneous generation of life. As a concept, spontaneous generation had fallen into disrepute by 1800, although even the work of Pasteur much later in the century, with his famous demonstration of 1859, would not entirely lay the problem to rest. While we can routinely think of sterilising food stuffs, in the nineteenth century, this was impossible and the technical problems of Pasteur's experiment resulted in some criticism of its efficacy in disproving spontaneous generation.

In Chambers line of reasoning, there is an increasing trail of complexity from the point of origin of life culminating in man. He accepts the idea of formation of new species as well as the extinction of old ones, which as he points out can be

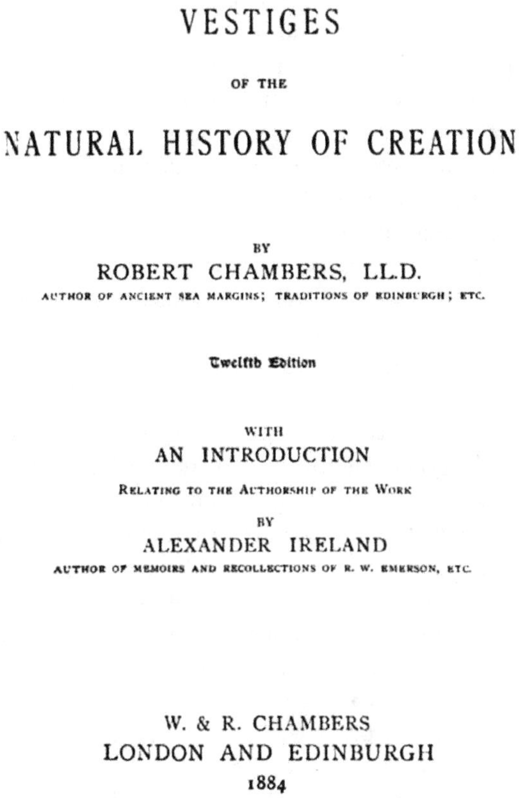

VESTIGES

OF THE

NATURAL HISTORY OF CREATION

BY
ROBERT CHAMBERS, LL.D.
AUTHOR OF ANCIENT SEA MARGINS; TRADITIONS OF EDINBURGH; ETC.

Twelfth Edition

WITH
AN INTRODUCTION
RELATING TO THE AUTHORSHIP OF THE WORK

BY
ALEXANDER IRELAND
AUTHOR OF MEMOIRS AND RECOLLECTIONS OF R. W. EMERSON, ETC.

W. & R. CHAMBERS
LONDON AND EDINBURGH
1884

Figure 3.12 Title page of Vestiges of Natural History of Creation by Robert Chambers. This is the twelfth edition, the first where authorship was acknowledged, previously it had been anonymous.

read in the fossil record. This process resulted in an understanding that there was no need for a continually working God to be actively creating new species. It was this content that angered the theologians, but was the very corner stone of its popularity as it encouraged a view that both science and theology can coexist.

Although it was a very popular work, the fact that it was controversial can be gauged by the overly critical response that it engendered in the scientific reviews. Much effort was put into criticizing the Lamarckian content. This is, however, unfair because he was a free thinker making a good attempt to evince a principle, but without detailed knowledge that the likes of Lyell would be expected to have. Vestiges was a good and largely successful attempt to put fossils into their true position as central to the age of the earth and later into their pivotal position in the story of evolution. Many of the contemporary reviewers seemed to be affronted and overly critical of someone making such momentous claims and ideas available to the general public. It was certainly filled with odd points, but hit the nail on the head in its portrayal of geology and fossils being instructive of the dynamic and changing system of the planet which has evolved over very long periods of time.

The work of Chambers was not, of course, the first description of the planet as being of immense age, but it was the first one that made a significant inroad into the popular perception of the biblical system. This work built on previous authors, the most significant of which was James Hutton. His publication in *Transactions of the Royal Society of Edinburgh* (1788) of *Theory of the Earth* was a wide ranging work on geology, leaning heavily upon his observations of geological formations in Scotland. He formulated the idea of uniformitarianism in which the features of the crust of the earth have been formed by natural processes over very long periods of time. It was the insistence on the very long time scale inherent in the work of Hutton that caused such consternation as it extended the age of the planet by observational science, magnitudes beyond the simple calculations of biblical scholars. Hutton also started the process of setting geology on a path towards a clearly defined science. The central idea that Hutton expounded was that the crust of the earth goes through a cyclic process of mountain building and erosion to sediment which would then form the strata which becomes visible. Hutton recognised by this that the youngest strata are at the top of sedimentary rocks. Although this created the rift between religion and science over the age of the Earth, it would only be the next century that would see Charles Lyell and Charles Darwin recognise the deep significance that the time scale involved would have for the interpretation of fossils and evolution.

It took a long time for fossils to be recognised as having originated from animal or plant material after the ascendency of the monotheistic religions. Before this time, it was easily accepted that unearthed bones, fossils, were the product of ancients or the bones of mythical beasts. The change to a religion that claimed a complete explanation of the creation of the sun and planets had no space for a history that was longer than it was possible to record. This limit to the interpretation of fossilised material carried on until it became impossible for the church to deny the evidence. Even though it became increasingly difficult for alternative explanations to cope with all the observations, the wish not to offend the religious sensibilities of either the hierarchy or the laity made general acceptance of geology and palaeontology a slow process. By the middle of the nineteenth century, the situation had changed and even grand attempts like *Omphalos* by Philip Henry Gosse in 1857 could not stem the tide that was changing the way that the physical world was being looked at. This was a time when a man seemed to be taking control of nature, no longer were horses the only way to move goods around the country, in fact perishable goods could now be moved by train over what had previously been inconceivable distances to market. The steam engine seemed to hold the promise of endless power for ships as well as factories and mines. Nature was yielding to humanity, and as a part of this an explanation of geological formations devoid of supernatural input seemed perfectly reasonable. There were, of course, those with misgivings, still uncertain about the implied atheism that denying the input of the divine would hint at. None the less, by the nineteenth century, fossils were centre to the idea of ancient life, epitomised by the birth of Mary Anning in 1799. Mary would become the most outstanding finder of fossils on the south coast of England, discovering many new and unusual species.

With the acceptance of fossils as part of the natural zoology of the past, they became important in the development of evolutionary theory, as well as having their own taxonomy. This development of apparent relationships between long extinct species moved on to an overall taxonomy of the extant and extinct, where the extinct species formed as much a part of the knowledge of the development of life as did the known living species.

References

Agricola, G. (1556). *De re Metallica*. (trans. 1912 H.C. Hoover and L.H. Hoover). London: Private Publication.

Burnet, T. (1691). *The Sacred Theory of the Earth*. Fleet Street, London: Lintot.

Carroll, L. (1865). *Alice's Adventures in Wonderland*. London: Macmillan.

Gesner, C. (1558). Historiae Animalium. Tiguri (Zurich).

Gesner, C. (1565). *De Rerum fossilium, lapidum et gemmarum maxime*. Tiguri (Zurich).

Gosse, P. (1857). *Omphalos: An Attempt to Untie the Geological Knot*. London: William Heineman.

Hooke, R. (1665). *Micrographia*. London: J. Martyn and J. Allestry, Printers to The Royal Society.

Hutton, J. (1788). Theory of the earth. *Transactions of the Royal Society of Edinburgh* 1: 209–304.

Lhwyd, E. (1699). *Lithophylacii Britannici Ichnographia*. London: Printed for Subscribers.

de Maillet, B. (1750). *Telliamed, or the World Explained*. (translation). London: Osborne.

Mercati, M. (1719). *Metalotheca Vatcana*. Rome.

Narborough, S.J. (1694). *An Account of Several Late Voyages and Discoveries to the South and North towards the Straits of Magellan, the South Seas and the Vast Tracts of Land Beyond*. London: St Pauls.

Plot, R. (1667). *The Natural History of Oxford-Shire*. London.

Ray, J. (1673). *Observations Topographical, Moral and Physiological Made in a Journey Through Part of the Low-Countries* Printed for John Martyn. London.

Ray, J. (1692). *Concerning the Dissolution of the World*. St Pauls, London: Samuel Smith.

Ray, J. (1693). *Three Physico-Theological Discourses*, Thirde (1713). St Pauls, London: Samuel Smith.

Rudwick, M.J.S. (1972). *The Meaning of Fossils*. London: Macdonald.

Steno, N. (1667). *Canis Carchariae Dissectum Caput*. Florentiae.

Steno, N. (1669). *De Solido Intra Solidum Naturaliter Contento Dissertationis Prodromus*. Florentiae.

White, G. (1789). *The Natural History and Antiquities of Selborne*. Recent edition. Penguin English Library (1977).

Woodward, J. (1695). *A Natural History of the Earth and Terrestrial Bodies, Especially Minerals*. St Pauls, London: R. Wilkin.

4

Reconstructing Animals from Fossils

Given the damaged skull of a long extinct species that no one has ever seen alive and upon being asked to describe, or at least imagine the living animal, how would it be approached? The natural reaction is to draw on those living species that we do know. For the earliest reconstructors, then, this was not a random activity, but based on what was known of anatomy from similar species. Before this systematic method took hold, there was a long period when reconstruction was based on a different set of criteria. If the fossilised bones were obviously not like anything known at the time, corresponding to nothing that was described from foreign travels, then imagination would be the first point of contact. As we have seen in Chapter 2, imagination and mythology can play off each other to create what to all intense and purposes is a real animal, with eventual claims appearing from individuals who say they have seen the animal in the flesh.

An often overlooked aspect of early illustrators of fossilised remains is that they were limited by the available techniques in printing. While the artist may have been more than able to create an exact likeness, by the time the image appeared in print, it was quite often several iterations old, gradually accumulating errors throughout the process. These errors would primarily take the form of changes in line thickness and ink spots.

It is not any surprise that there are very few illustrations of fossilised material as centre stage before the advent of printed material. Before the introduction of printed documents, the choice of subject for illustration was less factual and more religious, leaving little space or time for speculations regarding found material. When printing was introduced into Europe by Johannes Gutenberg (1398–1468), using moveable type, in about 1439, the situation regarding illustrated works changed completely. Gutenberg also introduced the use of oil-based inks, rather than the water-based inks which were commonly used by scribes and copyists. The change that printing made, besides availability of the written word to a wider audience, was that it became possible to insert illustrations. These were originally

Investigating Fossils: A History of Palaeontology, First Edition. Wilson J. Wall.
© 2021 John Wiley & Sons Ltd. Published 2021 by John Wiley & Sons Ltd.

woodcuts, which maintained their ascendency as the most common method of book illustration until the end of the sixteenth century.

Although woodcuts were used to illustrate the earliest books about fossils, such as *De Rerum fossilium, Lapidum et Gemmarum maxime*, by Conrad Gesner, published in 1565 (Figure 4.1), in western Europe, little was thought regarding the process of creating these woodcut images from original designs. This is a shame as although Gesner produced the first fossil illustrations, it was not him alone who was responsible for the finished work. At that time, the production of works including illustrations was a very skilled affair. We can get a better idea of the process and importance of the team who created these works by looking further afield at the production of woodcuts for purely artistic reasons.

In Japan, we find a well-documented system of woodcut making. Although the *Ukiyo-e* style that we shall look at did not properly appear until the second half of the seventeenth century. But similar to all such art forms, it developed from a

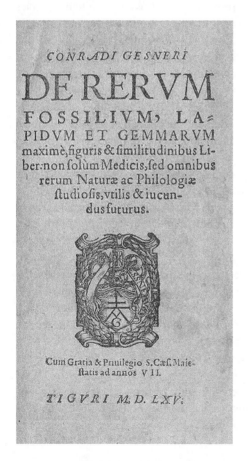

Figure 4.1   *The title page of* De Rerum fossilium, Lapidum et Gemmarum maxime, *by Conrad Gesner from 1565. An early book illustrating fossils with woodcuts.*

much earlier tradition. Although you may not be familiar with the idea of *Ukiyo-e,* you will most likely have seen reproductions of *The Great Wave* produced by Hokusai (1760–1849) around 1830 (Figure 4.2).

Colour prints such as these were recognised as collaborations between four people. These were the publisher, who commissioned the picture and coordinated the other members of the production team; the artist, often referred to as the designer; the block cutter and the printer. The printer is included because producing high-quality prints was a delicate and skilled task. In some ways, since paper was handmade, the papermaker could also be included in this list.

The artist would supply the drawing at the size it was to be reproduced to the block maker. This would be pasted face down on a block of wood with the next step being the entire responsibility of the wood cutter. The wood was usually fruit wood, such as cherry or pear, walnut or boxwood. As woodcuts developed in Europe, it became normal to cut the block on the end grain of wood. This was championed as a technique by Thomas Beswick and is normally called wood engraving to differentiate it from long grain woodcuts. In Japan, where cherry was the favoured wood, it would be the long grain side that was used as the hard tight grain would take accurate cutting quite well. This was also the side used by cutters working with Albrecht Dürer in the early sixteenth century. As can be appreciated, the artists original would be destroyed in cutting the block. It is interesting

Figure 4.2 The Great Wave *by Hokusai, this is one of the most highly reproduced woodcut illustrations ever produced. It was the first print in a series,* 36 views of Mount Fuji.

that the well-known illustration of an Indian rhinoceros dated 1515 by Dürer was reproduced elsewhere including *Historiae Animalium* by Conrad Gesner (Figure 4.3) which was published in four volumes between 1551 and 1558. It should also be recognised that text which was an integral part of the illustration would also have to be cut as reverse figures by the block maker. Other text not directly part of the illustration was printed using moveable type.

(a)

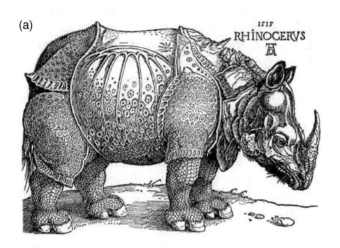

(b)

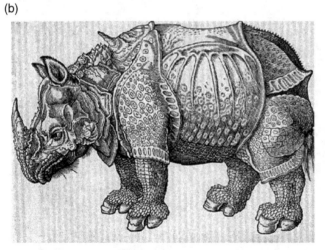

Figure 4.3 (a) *The original illustration by Albrecht Dürer of an Indian rhinoceros. Source:* Albrecht Dürer. (b) *A copy of the Indian Rhinoceros as used by Gesner in* Historiae Animalium Volume 1, 1551. *It was the way in which the copy was made that resulted in the image being reversed. The skill of Dürer is easily recognised as he never actually saw the animal. Source:* Gesner in Historiae Animalium Volume 1, 1551.

For palaeontology, this division of labour between skilled scientific artist and skilled block maker meant that there were some exceptionally clear and precise illustrations produced from which we can identify the fossil species. At this time, nearly all of the illustrated species were of shells or hard parts from species that were recognisably animal. They may not have been recognised as specific organisms, but they were nearly all shells or sharks teeth, the latter being described at the time as *glossopetrae,* literally 'tongue stones'. It was the use of woodcut illustrations of *glossopetrae* side by side with a contemporary shark that persuaded the scientific community that these items were in some way related. The very first example of this was by Gesner in *Historiae Animalium* Volume IIII of 1558 (Figure 3.2). In this woodcut, the largest part of the illustration is the *glossopetrae,* with the shark curved around it. Interestingly, the shark shows signs of having been dissected; more than that, it is possible to surmise that the shark is *Hexanchus.* This is based on the shark being shown with six gill openings, a relatively unusual situation. *Hexanchus* is a deep water species from the Mediterranean and north Atlantic, but whether Gesner actually saw the animal or worked from descriptions or sketches that were sent to him by corresponding scientists, we cannot be sure.

In many ways, these illustrations were *aide-memoires* for the reader, rather than accurate depictions of specimens. It allowed the curiousness to link the written description to the image and thence develop an understanding of the lines of reasoning that were propounded in the text. Much later, in 1667, Nicolas Steno (1638–1686) published *Canis Carchariae Dissectum Caput* (Steno 1667) which included a plate of a tongue stone, *glossopetrae,* and a sharks' head (Figure 3.11). This was interesting for many reasons, not least of which was that it was a metal plate engraving, rather than a woodcut. It was the first explicit statement that tongue stones were fossilised sharks teeth. This was also mentioned in a review of books in the *Philosophical Transaction* of the Royal Society for 1667. The image itself was not from a drawing by Steno, but had been made much earlier by Michele Mercati (1541–1593) although never used. The sharks' head illustration has an anthropomorphic aspect, although rather of the style of James Gilray. All of these illustrations were monochrome prints, with the greatest detail and clarity being found on the earliest prints. With time, the fine details of a woodcut become worn, but this occurs at such a rate that it requires comparison between the first and later prints for it to be manifest. This assumes that the print run was long enough for the wear to show up, which for many of these books it was not.

The move towards engraved printing plates allowed for far greater detail to be carried by the image. The development of engraving in Europe seems to have started simultaneously in both Italy and Germany. It was probably not coincidence that it originated in these two centres at about the same time as it was developed by goldsmiths, a major profession in both countries. The fundamental difference between engraved images and woodcuts is that while woodcuts are a relief printing technique, engraving is an incised technique. Engraving is referred to as being intaglio, referring to the incised image, which is then filled with ink. The surface of the engraved plate is then wiped clear of extraneous ink so that when paper is pressed down on the plate, it takes up ink from the incised cuts. This method of printing has the advantage of giving more detail as the plate is made of a harder material than wood, allowing for finer lines to be delineated. It is also this same resilience, the hardness, which reduces wear on the plate and increases the number of prints that can be made from each illustration. The widespread use of engraved illustrations did not really start until the end of the fifteenth century, developing later than woodcuts as a method of creating a reliable printed illustration.

After engraving had been invented and was being used in books, the process of etching was developed. This is a process in which a wax layer is used to cover a plate and a fine instrument then scratches the wax from the surface revealing the metal. Once the image is completed, the plate is washed in acid, which has access to the metal in specific areas and leaves the image permanently recorded as a very slight indentation, created by the action of the acid. A variation of etching is lithography. Invented by Alois Senefelder in 1796, this process uses a slab of lithographic limestone, which usually has a grain size of <0.004 mm (1/250 mm), which is coated in wax to proof it against the action of acids. This then has a design scratched through the wax to expose the limestone. The limestone is then acid-washed to partially dissolve the exposed surface, which becomes more ink absorbent so that the image can be transferred to a sheet of paper that is applied and pressed down onto the stone. The original source of lithographic limestone was the late Jurassic deposit at Solnhofen, this particular stone being easily split into thin plates. One of these blocks of stone from Solnhofen was split to reveal the fossilised skeleton of a part reptile and part bird. This became known as *Archaeopteryx lithographica*, reflecting the origin of the fossil being in lithographic limestone.

Most of the illustrations of fossilised material during this period were of shells rather than whole animals, and single bones were very rarely illustrated. There are several reasons for this, the primary one being that a shell, fossilised or not, is a recognisable object of aesthetic value. A secondary reason is that skeletal remains were still difficult to recognise and even more difficult to reconcile with the prevailing, mainly biblical, thoughts regarding the world and all the life on it. Also, of course, part of a bone from an unknown species does not generally make for an appealing image. The authors of the day would also be very aware that the illustrations needed to be attractive to the reader.

Given this tendency to illustrate fossil shells of all sorts, it is interesting to note that fossil coral and some parts of crinoids appear quite frequently in early illustrations of fossils. From this, it is possible to surmise that there was an understanding that these were copies, if not the actual bodies, of once extant animal forms. This does not mean that representations of complete animals were unknown before this time, simply that they were rare compared with shells. Of course, there were representations of griffins, which as we have seen, may or may not, have been influenced by fossil remains of ceratopsians, most notably *Protoceratops*. But for a documented representation of an animal influenced by a fossil find, we have a precise date of one of the very first.

Klagenfurt in southern Austria has a legend associated with its origin where two men killed the *Lindwurm*, a winged dragon, and founded the town. This folk-lore event occurred at an unspecified date in the past, but in 1593 Ulrich Vogelsang completed a statue of the animal in the form of a six-legged dragon. The palaeon-tological interest is that the head of the sculpture was based upon a skull that had been found in a local quarry in 1335 and which was believed to be the very skull of the killed *Lindwurm*. The sculpture itself is made from a single piece of chlorite schist. Being relatively soft, it is a common stone used in carving and has been used since prehistoric times for this. Strangely, about 39 years later, a statue of Hercules standing in front of the dragon was added to the original statue. Although this bears no resemblance to the original story, it does add some artistic balance to the statue. What makes this sculpture so significant is that although the overall image was based on myth and imagination, the head was based on the skull which was found over 250 years earlier. This skull had been kept, not only long enough to inspire the head of the statue but also long enough to be clearly identified and is still available to be seen in the local museum. It is now precisely identified as coming from a woolly rhinoceros, *Coelodonta antiquitatis*. This species appears first in the fossil record only about 350 000 years ago and disappears about 10 000 years ago. The fossils are relatively common, some being very well pre-served in permafrost and surface oil deposits and according to Hutchinson (1892), skulls of this species were occasionally dredged up from the Dogger Bank in the North Sea by trawlers. The sculpture of the *Lindwurm* was an exceptional produc-tion, and there would not be a similar attempt at three-dimensional representa-tions of fossil species for a long time. In the meantime, it was illustrations, mainly in books, that would take centre stage in helping people to understand the extinct flora and fauna found in the fossil record.

For most of the period from the first curiosity of fossils until the nineteenth century, representations of fossils were made on the page, rather than three-dimensional reconstructions. Before the advent of photography, also in the nine-teenth century, it was the skill of the artist and engraver which gave fossil illustrations their veracity. The quality of reproduction does vary quite widely, reflecting not only the range of skills of the illustrators but also the subjective assertion by the illustrator of exactly which features should be brought to the fore.

It is William Stukley who is credited with the first depiction of a *Plesiosaur,* in 1719 (Figure 4.4). This illustration appeared in *Transactions of the Royal Society* and although the engraving is difficult to interpret, this is not surprising. While recognising and drawing a recognisable structure such as a shell is quite straight-forward, trying to make sense of a collapsed and crushed skeleton, with parts missing, is a very difficult task. At the time, this would be especially difficult because there would be no clear understanding of exactly what this barely recog-nisable pile of bones represented. Like many of the scientific reports of the day, space was not so restrained as it became in the second half of the twentieth cen-tury; consequently, we have more background information regarding the find than we might have done.

The slab of rock containing the fossil was given as roughly 3ft×2ft 2in. (c 91.5cm×66cm) and was being used as a loading stone for standing on, at the well of the Elston parsonage. It had been there for as long as anyone could remember until, when the stone was lifted, it was discovered that the underside contained the embedded fossil. The skeleton (spelt *sceleton* by Stukley using the modern Latin form, although *skeleton* had already been in use for nearly 150 years) was incomplete.

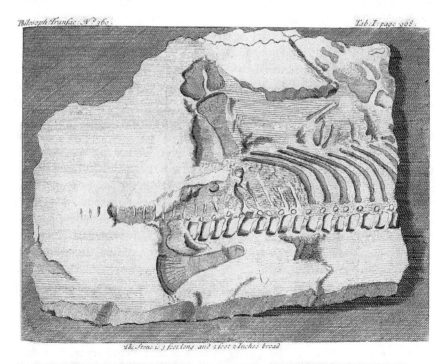

Figure 4.4 *The slab of rock containing the fossilised skeleton described by Stukley, now known to be* Plesiosaurus dolichodeirus, *one of the long-necked Jurassic marine reptiles.*

Both the block with the continuation of the fossil and the corresponding slab which fitted with it before the rock was longitudinally split had been lost in antiquity. The block was brought to the attention of Stukley by Robert Darwin, the great grandfather of Charles Darwin. Robert Darwin considered that it might have been human remains, as was the prevailing view of the parishioners of Elston. Recognising the importance and curiosity value of the slab, the Parson of Elston, John Smith, had displayed the fossil in the parsonage garden, where it gained a wider audience. Stukley was quick to realise that this was not a human skeleton, but by his calculation something like either a crocodile or porpoise. This was a very well reasoned deduction as we now recognise that this specimen is actually *Plesiosaurus dolichodeirus,* one of the long-necked Jurassic marine reptiles. Although this is a long way from either a crocodile (Reptilia) or porpoise (Mammalia), they are both adapted to an aquatic lifestyle.

Perhaps, the most significant part of the work by Stukley is that he was trying to determine the taxonomic position of an unknown species based only on the incomplete fossilised remains, as found in Nottinghamshire. There was no zoological point of reference at this time for Stukley to work from, other than that these were not human remains, remains with which, as a medical doctor, he would have been familiar. The work of Stukley started the wider process of illustrating partial fossil remains. Whenever they were found, confusion as to what they were remained as a constant reminder that there was little or no context for these zoological fossils. The natural assumption was that they were related to living species, so they would inevitably be put into the current taxonomy as it was understood at the time.

A prime example of fitting a fossil into a current taxonomy is found only a few years after the work of Stukley. In 1726, Johann Scheuchzer described and illustrated a Miocene fossil from Ohningen in Germany as *Homo Diluvii Testis,* thinking it was a partial human skeleton that had been crushed in the flood. Interestingly, this was picked up and repeated, both in print and by illustration, by Louis Bourguet who published *Traite des Petrifications* in 1742, with what is thought to be a copy of Scheuchzer's original engraving with additions. The copy is of a lower standard than the original (Figure 4.5). To a modern eye, the differences between the fossil and human remains may seem obvious, even more so to a modern biologist used to the enormous range of images of species from around the world. The illustration clearly shows an amphibian. It was only much later, in 1822, that Cuvier suggested that it was in fact a salamander, an idea that was recognised as correct when Cuvier removed more of the supporting rock to reveal the forelimbs of the salamander. This is one of many cases which demonstrate the detailed understanding of anatomy and biology that George Cuvier (1769–1832) developed.

In 1749, a remarkable work, *Protogaea*, was published posthumously by Gottfreid Wilhelm Leibniz (1646–1716). This had been written between 1691 and

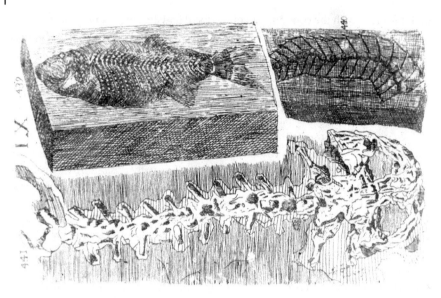

Figure 4.5 *The copy of* Homo Diluvii Testis *published by Louis Bourguet in 1742. The original by Scheuchzer had more detail, while the copy contained an additional two images of fish. Source:* Bourguet (1742)/Antoine-Claude Briasson/Public domain.

1693 but was not published for many years. This work clearly stated that fossils were not the act of the 'Great Architect' using animals as models for rock formations. This had been claimed by Athanasius Kircher (1602–1680) who also said that large fossilised bones came from a race of giant humans. Leibniz would have no truck with explanations where a hole was plugged by introducing a theistic body. Consequently, Leibniz clearly stated that *Glossopetrae* did not just look like shark teeth, they were shark's teeth. Similarly, fish preserved in slate were exactly that, not a miraculous deposition. Leibniz did not always hit the mark, however. One of the examples that he describes is the Quedlinburg Unicorn. This was a collection of bones that were discovered in 1663 and then constructed into a supposed unicorn, which were taken as face value as being a unicorn by Leibniz (1749) (Figure 4.6). The skeleton was the subject of an illustration that is a very good reproduction of the skeleton that is still on display in the Natural History Museum, Magdeburg. This illustration, like all the illustrations in *Protogaea,* are woodcuts. The long-term influence of *Protogaea* was surprisingly limited, considering the perceptive assessment of fossilised material. *Protogaea* was not translated into English until 2008, which left it languishing through the nineteenth and twentieth centuries as the habit of reading Latin texts became less commonplace. Primarily, however, it may have been the lack of any sense of geological time that left the text somehow disconnected to later critical thinking. Leibniz was writing

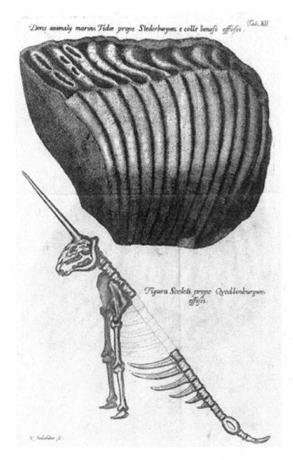

Figure 4.6 *An illustration of the Quedlinburg Unicorn, the animal skeleton assumed to be real by Leibnitz.*

at a period when biblical time was still generally recognised as correct, so while he did not deliberately condone it, there was never any sense that it should be contradicted. Although his ideas regarding fossils were sound and meaningful, this was not embraced wholeheartedly by his contemporaries.

It was around the middle of the eighteenth century that saw a change take place in the perception of fossils amongst the scientific community. This was associated with the work of Georges-Louis Leclerc, le Comte de Buffon (1707–1788) (Figure 4.7), generally just referred to as Buffon. He was an extraordinarily perceptive naturalist and by 1749 had produced a volume on the Earth, which was to form part of his *Natural History* of 36 volumes. In this volume, he described mechanisms that could be used to explain marine fossils being found on land,

Figure 4.7 *Portrait of Georges-Louis Leclerc, le Comte de Buffon (1707–1788) by François-Hubert Drouais in 1753. Musee Buffon, Montbard. Source:* Musee Buffon, Montbard.

explaining within a biblical time scale the events taking place in a more compliant and mobile planet surface.

Although the considerations of Buffon were considerably modified by the work of Leibniz when considering the cooling of the planet, Buffon was trying to calculate the age of the earth by scaling up from the cooling of metal balls. Although this gave a value of tens of thousands of years, Buffon was convinced that it would take a time scale measured in millions of years to explain the observed geological structures and most specifically the deposition of the observed strata. While Buffon was keen to explain observations as best he could, the interpretation of specific fossils was not approached in the same way. Although he was a significant naturalist, Buffon did not significantly improve on previous attempts to interpret fossils and fossilised remains. It was another French scientist, Cuvier, who would make palaeontology not only a proper discipline, but a part of biology as well. This was not something that had really been considered before, it was a philosophical problem because if these ostensibly extinct forms were truly part of biology, then the current ideas of creation were either incorrect, or the Church was, because extinction did not form part of the Great Plan as understood by the Church. It would have been realised at the time that such questions were best left unasked as they threatened the underpinnings of the establishment. While Buffon was the pre-eminent influence on Pre-revolutionary French natural history, his influence evaporated with the revolutionary zeal that openly rejected his stance as being both simultaneously elitist and disdainful of systems and methods that were seen as essential for developing a true knowledge of the natural world. These were the scientific ideas of observation and interpretation without preconception.

Into this arena of political turmoil and religious infallibility, there came a naturalist and anatomist of outstanding ability, Georges Cuvier (1769–1832). This was a time of turmoil, and yet when Cuvier gave a paper in Paris regarding the comparative anatomy of the known living and fossil elephants, the implications of his paper were instantly recognised – extinction was a real phenomenon. Delivered in 1796, acceptance of the ideas was probably aided by it being a time of change, both social and intellectual. Revolution was in the air; the First Republic had been declared in September 1792 and would last until 1804 when the First Empire was declared under Napoleon.

Such was the social upheaval that for about 12 years between 1793 and 1805, there existed the somewhat eccentric French Republic Calendar. This was designed to remove from the date all hints of elitism, which meant removing all religious and royalist references, so all the names of the months were altered as were the names of the days of the week, of which there were now 10, only one of which was a rest day. Although in time it would cause problems of seasonal drift, each month was made up of 3 weeks of 10 days each. It was into this period of change that Cuvier began to imply and then propound ideas of extinction, which as we have seen, to a strict adherent of biblical religion implies God's fallibility.

Cuvier was an exponent and advocate of Linnaean taxonomy, extending it to include the level of the phylum. Perhaps more importantly in an act of reconciliation between the living and fossil species, he included fossil forms into a taxonomic system which had been originally only devised for living species. This method of classification was devised by the Swede, Carl Linneaus (1707–1832) who, after ennoblement in 1761, became Carl von Linne. His major publication was *Systema Naturae*, where his name appears as Caroli Linnaei (1758) (Figure 4.8). This was the first systematic use of a binomial naming system in a book. Although the first edition was produced in 1735, it was the tenth edition of 1758 that is seen as the true origin of biological nomenclature. In this book, he did not just expound the system as an idea, but gave examples of about 6000 plants and 42 500 animals. Although this may seem like a large number, it should be remembered that we now know of well over 80 000 living species of molluscs as well as at

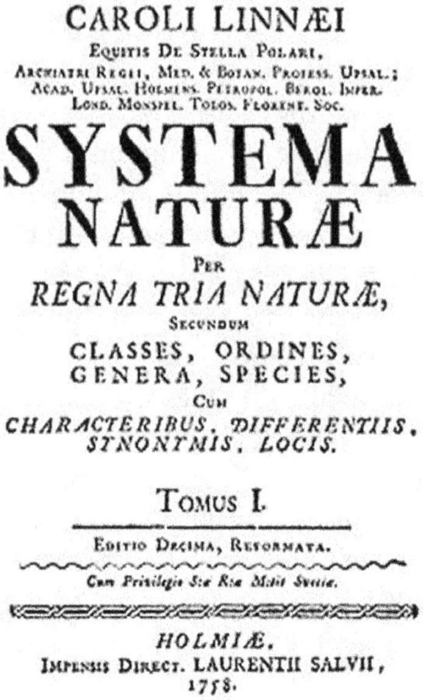

CAROLI LINNÆI

Equitis De Stella Polari,
Archiatri Regii, Med. & Botan. Profess. Upsal.;
Acad. Upsal. Holmens. Petropol. Berol. Imper.
Lond. Monspel. Tolos. Florent. Soc.

SYSTEMA
NATURÆ

Per
REGNA TRIA NATURÆ,

Secundum

CLASSES, ORDINES,
GENERA, SPECIES,

Cum

*CHARACTERIBUS, DIFFERENTIIS,
SYNONYMIS, LOCIS.*

Tomus I.

Editio Decima, Reformata.

Cum Privilegio S:æ R:æ M:tis Sveciæ.

HOLMIÆ,
Impensis Direct. LAURENTII SALVII,
1758.

Figure 4.8 *Title page of* Systema Naturae *from the tenth edition of 1758, the edition which marks the start of modern zoological nomenclature. Source:* Carl Linnaeus,https://en.wikipedia.org/wiki/Systema_Naturae#/media/File:Linnaeus1758-title-page.jpg,Public Domain.

least 35 000 fossil species. His catalogue became larger with each edition, as would be expected with the development of a robust system designed to be able to catalogue every species, living of dead, although his use of sexual organs in plants as the primary taxonomic feature has caused some problems in the past. His system had three Kingdoms, which were Animal, Vegetable and Mineral. It was without doubt pivotal in being able to understand the relationships between taxonomic groups, but it is the principle that has stood the test of time, rather than the structure of the system as originally laid down by Linnaeus. It is interesting that after the death of Linnaeus, his collection was in the hands of his wife and his daughter, but suffered many years of neglect and had begun to decay. In 1784, the collection was bought from the family by James Edward Smith and then shipped to London. The collection was used for research but the casual attitude towards it resulted in both losses and editions. It was eventually to become the kernel for the foundation of the Linnaean Society in 1788 by James Smith.

Cuvier had a philosophical attitude to natural history and most specifically anatomy, which would stand him in good stead when he worked on fossil material. This philosophy was based on the works of Aristotle and the reductionist views of the physical world expressed in the burgeoning industrial and social revolutions, where anything could be understood and described in fundamental terms. Just like the planetary systems, biological organisms were considered to be like machines, and like machines, could be reduced to the fundamental components of physics and chemistry. When looking at organisms, the metaphor of machines was pushed even further. Using the machine analogy to demonstrate that all the parts of a machine were required to work for the totality to function correctly, so all the structural components of an organism were required to work correctly for the organism to function fully. Taken to its logical conclusion, it would mean that there was very little available flexibility in an organism which could allow for evolutionary change as a single change would invalidate the overall function. So species were not individually constructed products, but real units of biology only capable of trivial variations in structure without compromising the integrity of the organism.

Cuvier's work for which he is best known, and would be pivotal in developing his skills in palaeontology, was based around two principles. The first of these we now see as self evident, it refers to the style of existence that an animal holds; carnivores have cutting teeth and strength to subdue prey. The second idea has a more nebulous feel about it, but has a greater significance. The characteristics of an organism as a whole can be used to classify it and its zoological connections, but there is a hierarchy within a group whereby some characters are more powerful in their influence on the classification and the clade to which it belongs. These ideas would take considerable investigation by Cuvier before they could be explicitly stated, but some notable successes using these ideas in the investigation of fossilised material would help the reputation of Cuvier.

In 1796, Cuvier published his paper demonstrating that African and Indian elephants are different species and mammoths were different to both of them, just from the study of skeletons (Cuvier 1796a). In the same year, he studied a large fossil skeleton that had been found in Paraguay. This was of considerable significance in the development of palaeontology as it marked a point at which comparative anatomy became a significant tool in understanding animal remains (Cuvier 1796b). It started when Cuvier was sent a series of unpublished engravings upon which he was asked to speculate. By this time, the skeleton from Paraguay was in Madrid and it was obvious that this was a very large animal. Nonetheless, Cuvier was sure, correctly, that this was the skeleton of a species which was both extinct and new to science. But more than this, even with the massive difference in scale, Cuvier put it in the same suborder as the sloths (Order: Pilosa, Suborder: Folivora, of which there are five families; three extinct, with all living species in two families). This was the first time that such a technique had been used in palaeontology and the result was most convincing (Figure 4.9). This was superbly demonstrated by the illustration in the paper where for ease of comparison, Cuvier had changed the scale so that the skulls, two of extant species and the extinct form, were the same size. Cuvier named the ground sloth as *Megatherium,* the name by which we know it (Figure 4.10a, b). Some of these species seem to have survived until quite recently, certainly to have been cohabiting

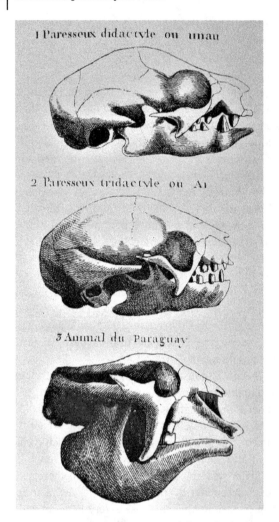

Figure 4.9 *Cuviers's comparison of skulls of living sloths with* Megatherium. *They are all adjusted for size to make anatomical comparisons easier. This was the first time such a comparison had been made.* Magasin Encyclopédique, *1, 1796, pp. 303–310. Source: Wilson J. Wall.*

with man, so there are a number of very well-preserved skeletons of these species from South America.

Having used his skill with such great affect, it was hardly surprising that with time, his fame would extend beyond palaeontology reflecting the ever burgeoning interest in all things scientific. This has no better demonstration than in the story written by Sir Arthur Conan Doyle, published in *Strand Magazine* in November 1891 with the title *The Five Orange Pips*.

(a)   (b)

GREAT GROUND-SLOTH OF SOUTH AMERICA, MEGATHERIUM AMERICANUM.
PLATE XVII.   Length 18 feet

Figure 4.10 (a) Megatherium, *as illustrated in 1892 by J. Smit in* Extinct Monsters *by H·N Hutchinson. It is captioned as 18 ft, about 5.5 m. The similarity to living sloths can easily be seen in this representation.* (b) Megatherium *skeleton from the Natural History Museum, London. This is made from casts of bones from several different skeletons and has been on display since it was constructed in 1848, influencing the reconstructions as shown in* Figure 4.10. *Source:* Wilson J. Wall.

> As Cuvier could correctly describe a whole animal by the contemplation of a single bone, so the observer who has thoroughly understood one link in a series of incidents should be able to accurately state all the other ones, both before and after.

Cuvier started to look at further examples of extinct species, many of which had been discussed at great length since the eighteenth century. Although focussed on what the fossils were, there was no conclusion regarding extinction, or even the nature of the species that were represented. Cuvier worked on fossils from illustrations provided to him which detailed found material. Using this technique, he described the first Pterosaur, a Pterodactyl, from a fossil found in Bavaria. The ability to make these significant discoveries and pronouncements was aided by a very high quality of illustration. Such standards were not surprising as well-trained geologists and palaeontologists were also highly competent artists and accurate recorders of the visible material; there was at the time no other method of recording an image other than drawing it. There is an advantage which a drawing has over a photographic record, by retaining everything that is seen as relevant in focus and that which is not regarded as important either ignored or rendered as hidden background.

For palaeontology in the nineteenth century, Cuvier changed one thing above all else; he legitimised the representation of fossil remains as living animals in their own right. Until this time, fossils had been seen as representing species unknown, or new to science, but few if any attempts were made to represent them in life. With the work of Cuvier, it became an acceptable idea that with care and knowledge, the jumble of bones could be reconstructed into an animal of recognisable pedigree, placed within a Linnaean taxonomy. Cuvier himself never produced illustrations of complete animals based on fossil skeletons for publication. It quickly, and quite logically, came about that since the skeleton was reconstructed, why not clothe the result in muscle and skin? It would seem that in a private note that Cuvier produced, there are examples where muscles had been sketched onto the illustrated skeleton, although they went no further than private sketches.

Before the advent of model making, which would become common towards the middle of the nineteenth century, it was the drawn illustration which was used as the starting point in depicting the fossil animal as it would be in life. The drawn image would also remain the mainstay of palaeontological record keeping long after photography became available. This was mainly due to the interpretive value of a drawing but also the limited accuracy of registration that was available on early film stock.

Photography was quickly taken up in scientific disciplines where an image was more straightforward in content than would be expected to be found in biological materials. One such science, of course, is astronomy where registering details of structure was not as important as registering relative positions in the sky of visual phenomenon.

Although photo-active chemicals had been known for a long time, controlling them was difficult and creating an image with them almost impossible. Creation of a flat plate in a focal plane was not a problem, but forming a reliable chemistry and a lens was. The first photographic systems needed very long exposures, sometimes being measured in hours. This was the case with images created using the system devised by Daguerre in France, the Daguerreotype, which also had the disadvantage of not being reproducible in that the image created was the final positive image. It was Fox-Talbot who devised a system which created a re-printable negative at about the same time that Daguerre was working. The first Fox-Talbot print from a negative was of the Oriel window of Lacock Abbey and is dated 1835. This was a low-resolution image, and Daguerreotypes remained preeminent until around 1850 when wet plate collodion photography was introduced. This still needed a long exposure and as the name implies, the plate was wet, so it was inclined to dry out if the exposure was too long. In 1856, a dry plate system was patented and quickly replaced the wet plate method, with the advantage over the wet system that it did not dry out on long exposures and the advantage over Daguerreotypes of producing a negative that could produce repeatable prints which could be circulated amongst interested parties. Although Daguerreotypes were going to be superseded around 1837, Daguerre took an image of fossil shells. Nonetheless, it was only by the end of the nineteenth century that photography became both routinely available and reliable for use in museums and in the field. Even then, it was not going to be a tool for the restoration of animals in life that was still the province of the skilled illustrator.

It was when complete, or partially complete, skeletons were first found and were starting to be put together that palaeontologists began to speculate as to what these animals looked like in life. Since most palaeontologists of the day were also skilled illustrators of their finds, they started to produce images that were speculations on antediluvian life. Fuelling this development was the work of Mary Anning (1799–1847). She was a collector of great skill, being aware of the high productivity of the fossil bearing strata of the blue lias formation that made up the cliffs of Dorset on the south coast of England.

For the Anning family, collecting fossils was a business which Mary learnt from her father, Richard. During the nineteenth century, tourism was becoming far more normal among the burgeoning middle classes as annual holidays were being taken. Richard Anning was a cabinet maker by trade, but started dealing in fossils when it became apparent that there was a market to be tapped. Mary learnt the art of fossil hunting after winter storms had brought down sections of the cliffs to reveal another crop of fossils. This was not then, and is still not now, carried out without risk; while out collecting in 1833 a landslip killed her dog, Tray, and narrowly missed causing her injury. Her dog was so well known as her accompanying familiar that it appears in a painting of her posing in her collecting clothes in front of the cliffs at Lyme Regis (Figure 4.11).

It is said that when she was 10 years old, Mary was paid half a crown for an ammonite which came from the beach. To put this into perspective, before February 1971, the UK had a non-decimal currency, based on no particular numerical system. Half a crown was 1/8th of a pound, there being four crowns to the pound, so a crown was 25 decimal pence. This was, however, a very simple part of the currency. There were 12 shillings to a pound (the half crown was 2 shillings and 6 pence, written 2/6 or 2s6d); each shilling was 12d (p for pence is a decimal notation), so there were 240 pennies in a pound. And a Guinea was 21 s or £1.1 s. For Mary, half a crown was a reasonable sum for a found object as a few years later, in 1834, according to the Poor Law Commission the day rate for a labourer was 10s 5½d.

While Anning was the most significant professional collector of the nineteenth century and her finds were of quite extraordinary importance, like the first correctly identified *Ichthyosaur*, the representation and presentation of her findings were the province of her personal friend Henry De la Beche (1796–1855) and William Conybeare (1787–1857).

The *Ichthyosaur* was originally discovered by Joseph Anning, the brother of Mary, who was unable to complete the excavation as he was indentured to an upholsterer and did not have the free time. Mary completed the excavation and

Figure 4.11 *Portrait of Mary Anning with her dog Tray. In the background are the cliffs of the south coast of England where Mary found her fossils. Probably painted during the 1840s.*

reassembled the skeleton which was 17 ft in length (5.18 m). This completed work was sold to a London collector for £23 and displayed by him until 1819 when it was sold to the British Museum for £24 (Figure 4.12). In 1823, Mary excavated a complete *Pleisosaurus*, for which she received £110. There was a constant series of discoveries and repeat findings, including the first pterosaur found in Britain. All of the findings were documented and many communicated to the Geological Society, but the most significant result of this high-density fossil deposit was for it to be recognised as an ecosystem. Although the term ecology was coined by Ernst Haeckel and first seen in English in 1873, the concept was much older.

Although ideas of interlinked organisms had been around for a long time, it is not regarded as having been formalised until Johannes Eugene Warming (1841–1924) wrote what is regarded as the first text book on plant ecology, and is regarded as formalising ecology and its study.

When the finds of Mary Anning were recognised as the important fossils they were, it was no surprise that palaeontologists of the day also recognised that instead of a bed of shells, they were looking into the past at a preserved snapshot

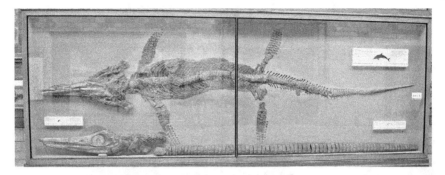

Figure 4.12 *An Ichthyosaur fossil excavated by Mary Anning and still on display at the Natural History Museum, London. Source:* Gary Todd, https://commons.wikimedia.org/wiki/File:Temnodontosaurus_skeletons.jpg#/media/File:Temnodontosaurus_skeletons.jpg, CC0.

of an ecosystem. This was epitomised by the painting by De la Beche which was titled *Duria Antiquior* or *A More Ancient Dorset* (Figure 4.13). This important depiction of fossils 'in life' was painted in 1830 and was based on the fossils which Anning had unearthed over the years. The original work is now housed in the National Museum of Wales, De la Beche having moved much of his geological collection to south Wales to be in easy reach of the coalfields of that area. This illustration was not only the first image depicting the interactions of fossil species, but it was also the first thoughtful attempt to depict fossil species alive, based on available scientific knowledge.

This knowledge came about from Anning observing that there was fossil material enclosed within the stomach of some of the larger fossil reptiles, inviting a direct assumption of a carnivorous diet. This led William Buckland (1784–1856) to write a description of the perceived food chain of the Lias, which in turn influenced De la Beche to produce his overly carnivorous illustration. De La Beche commissioned the well-known professional artist and print maker George Scharf (1788–1860) to produce lithographs based upon *Duria Antiquior* which were sold to raise money for Anning. These were sold privately for £2 10s, the equivalent of 12 days work for a skilled tradesman. With failing health, Mary was awarded an anuity raised by the British Association for the Advancement of Science and the Geological Society. Mary Anning is buried in the Church Yard of St Michaels Church in Lyme Regis, the grave stone still being tended (Figure 4.14). In recognition of her pioneering work, the Geological Society commissioned a stained glass window for St Michaels Church and a short distance away is Anning Road.

There are many features found in *Duria Antiquior* which shows the limits to knowledge of these animals at the time. Interestingly, the engraving made by Scharf from the original watercolour does have some significant differences. The original by De la Beche shows a range of organisms as found in the Dorset cliffs. They are mostly predatory, with some teleost fish being target prey. The two most

Figure 4.13   Duria Antiquior or A More Ancient Dorset, *painted by Henry de la Beche (1796–1855) in 1830. The original is in the National Museum of Wales, though the image has been much reproduced from lithographs of the original. Source:* Henry De la Beche, https://commons.wikimedia.org/wiki/File:Duria_Antiquior.jpg#/media/File:Duria_Antiquior.jpg, Public Domain.

Figure 4.14   *The tended grave of Mary Anning and her brother, Joseph in St Michaels Church, Lyme Regis.*

obvious points of interest from a zoological point of view involve the various Ichthyosaurs which are represented. Being reptiles, the ichthyosaur has a long tail, and since there are no extant marine reptiles of this type, it was reasonable to assume that the tail was straight, much like a lizard, with no caudal fin, also with no dorsal fin. We now know, from fossils found at Wurtemburg in 1892, that the vertebrae extend into and down the bottom edge of a fish-like upright caudal fin, had a dorsal fin and that the ichthyosaurs occupied a similar niche to modern dolphins. Rather more interesting is the speculation which resulted in De la Beche illustrating his Ichthyosaurs as venting from their nostrils in the same way that at the time, whales were thought to do. There is, however, a major difference in the position of the nostrils between whales (high on the head) and Ichthyosaurs (in front of the eyes and positioned like a bird).

By the time that Scharf had refined and modified the original work of Beche for publication, there had been some changes in detail. While the original work contained a plesiosaur snapping at the tail of a crocodile on the bank of the sea, in the lithograph, the plesiosaur has been repositioned to be attacking a submerged turtle. Scharf also made an alteration to the prey of the Ichthyosaurs. Beche had drawn the prey fish being taken from behind, but Scharf decided it was more realistic to have them being captured head-on. In the Beche painting, there is a fish which appears to be about to prey on another fish, while Scharf has completely changed this so that the fish is consuming a Eurypterid.

There was a lasting legacy of *Duria antiquior,* mainly through the prints by Scarf, but originating with the watercolour by De la Beche. Although the image may have been selectively chosen from the complicated multi-species original, there are cases where the illustrations are obviously inspired by the Scharf original. Two examples of this, both containing a Plesiosaur and an Ichthyosaur in the same pose, although rearranged in position, can be found in *Geology for Teachers, Classes and Private Students*, published in 1860 in America and *La Terre avant le deluge*, by Louis Figuier, Paris 1863 (Figure 4.15a, b). The key to their inspiration comes from the venting Ichthyosaur which is centre stage in both images. The influence of the idea of a venting Ichthyosaur seems to have extended into other species. In 1872, William Webb wrote *Buffalo Land*, with illustrations by Henry Worrall, one of which contained a Mosasaur that was also venting (Figure 4.16). The Mosasaur is labelled as being *Liodon*, which hangs on as a dubious genus.

This influence of illustrations on the depiction of ancient fossil species has been a running process more or less since the first publication of *Duria antiquior*. The illustrations with the most authority scattering their influence quite widely.

Although the venting of Ichthyosaurs through their nostrils in the same way as whales were thought to was based only on the similarity of Ichthyosaurs to dolphins. It should be remembered that whales do not actually spout water that was an easily made mistake, especially in cold weather. There may have been a small humorous conceit involved in the illustration as well. De la Beche was a skilled artist as well as a draughtsman and had produced satirical cartoons and drawings

(a)

(b)

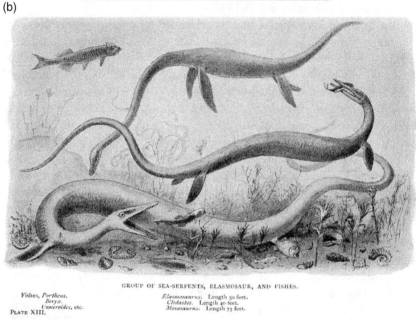

GROUP OF SEA-SERPENTS, ELASMOSAUR, AND FISHES.

Fishes, *Portheus.*
Beryx.
Osmeroides, etc.
PLATE XIII.

Elasmosaurus. Length 50 feet.
Clidastes. Length 40 feet.
Mosasaurus. Length 75 feet.

Figure 4.15 (a) La Terre avant le deluge *(1863) by Louis Figuier. The spouting Ichthyoaur is speculative, influenced by* Duria antiquior. (b) *Plesiosaurs as depicted in* La Terre avant le deluge *(1863) by Louis Figuier. Source:* Louis Figuier.

of his colleagues, such as *A Coprolitic Vision*, satirising William Buckland's apparent fascination with coprolites and *Awful Changes* on the possibilities of recurrent evolution. This raises the possibility that Ichthyosaurs were shown spouting as a touch of humour, certainly in the original watercolour the water spouts seem to have been casually added.

The depiction of palaeoecology in print became quite a common method of illustrating fossil animals as they would have been in life. The interpretation of

Figure 4.16 *From* Buffalo Land *by William Webb, illustration by Henry Worrall. The spouting form is* Liodon proriger (*now* Tylosaurus) *a member of the Mosasauridae.*

fossil remains is, after all, just that, an interpretation since it is rare for any soft parts to be preserved well enough to indicate an appearance in life. Bone masses would also normally be jumbled, so an exact relationship between fossil and life can be extremely difficult to correlate. A good example of this is the *Iguanodon*, only the second fossil reptile to be named after *Megalosaurus*.

The story of *Iguanodon* starts with the discovery by Gideon Mantell (1790–1852) of teeth and parts of an unknown skeleton (Mantel 1825). This collection of parts was not reconstructed, but interpreted as it was found, embedded in stone. The result of this was a quadruped with a small nose horn. The horn was the result of interpretation of the presence of an additional bone of unknown origin within the fossilised remains. Although not reconstructed, there is a sketch made by Mantell of his idea of the complete animal skeleton, which shows a quadruped with a long and flexible tail and a small horn on its nose. The original fossil is now in the Natural History Museum in London and has been reclassified as *Mantellodon*. Interestingly, Cuvier thought this was most probably a herbivorous reptile, although for many years was most definitely not how *Iguanodon* was portrayed. In 1865, an English translation appeared of *La terre avant le deluge* with the title *The World Before the Deluge* in which there was an illustration of a fight between an *Iguanodon* and *Megalosaurus*. In this illustration, they both look very dog-like and the *Iguanodon* has a nose horn, they also take up the form of a sort of two animal ouroboros (Figure 4.17).

Clarity in skeletal reconstruction of *Iguanodon* was slow in coming, even after what was recognised at the time as very significant finds in 1878. The find was of

XXI —Ideal scene in the Lower Cretaceous Period, with Iguanodon and Megalosaurus.

Figure 4.17 La terre avant le deluge, Iguanodon *and* megalosaurus *in mortal combat, although we know that* Iguanodon *was herbivorous and such a conflict would be one sided as the* Megalosaur *was a meat eating predator.*

29 (originally thought to be only 23) *Iguanodon* skeletons in a colliery at Bernissart in Belgium. Although the finds were associated with coal measures, they were not actually part of them. Folding of the coal seam had resulted in deep chasms which had been filled in overtime with what was described as chalk, it was in these deposits that the fossils were found. Of the 29 specimens recognised in 1878, by 1883 15 had been chiselled out and 2 complete skeletons reconstructed. At the time of the reconstruction of the skeletons at the Royal Museum of Natural History in Brussels, it was thought that the correct stance, due to the much more powerful hind legs, was along the lines of a wallaby or kangaroo, standing upright when on shore (Figure 4.18). The size and structure of the tail also implied that the reptile spent a great deal of time in water, where the tail was used for propulsion. These assumptions were quite reasonable because it seemed likely that these ancient species would move and act much as currently extant species of similar shape would.

By the middle of the twentieth century, it was generally recognised that the strangely bent tail of the wallaby model was incorrect and the true form was more likely standing on its hind legs with the tail held out behind as a counter balance, the front legs being used to balance when necessary, but not generally for ambulation.

While the various different ideas on the reconstruction went from four footed ambulation to upright wallaby-style and onto essentially a bipedal herbivore, one

(a)

A GIGANTIC DINOSAUR, IGUANODON BERNISSARTENSIS.
Length about 30 feet.

PLATE VII.

(b)   (c)

Figure 4.18 (a) *The outdated wallaby model of Iguanodon where the stance was tripod with the tail being used as an additional limb. The length of 30 ft is a little over nine metres. Source:* From Hutchinson, H.N. (1892) Extinct Monsters. D. Appleton and Company, New York. (b) *Modern interpretation of the stance of* Iguanodon. *This specimen from the Natural History Museum, London. Source:* Wilson J. Wall. (c) T*he renovated sculptures of* Iguanodon *residing in the park at Crystal Palace in south London. Note the horn on the nose as suggested by Mantell.*

of the lasting images is that of the rhino-like animal with a nose horn. This was an image promulgated in the nineteenth century in many popular works, but also as a sculpture in at Crystal Palace in south London. In 1852, Benjamin Waterhouse Hawkins was commissioned to produce 33 life-sized models in 15 genera, not all of them dinosaurs, but all of them extinct. Although the scientific advisor was Richard Owen, the *Iguanodon* retained the small horn on the nose, even though Owen had concluded that it was in fact a claw. The models are still in the park at Crystal Palace, although the glass house from which the park takes its name is no longer there. As the models are now Grade 1 listed, they have to be maintained as they were originally constructed, regardless of zoological accuracy.

While *Iguanodon* skeletons were reconstructed in Brussels and modelled in London, these were not the first attempt to represent dinosaurs as in life. That accolade goes to a cast of an almost complete skeleton of a Hadrosaur constructed in 1868 by Benjamin Waterhouse Hawkins and displayed at Philadelphia Museum.

It is not just dinosaurs that have been reconstructed in strange ways. Mammoths have also been reconstructed with errors, although these errors seem to be confined to the tusks since the skeleton follows the same plan as modern elephants. There was a particularly interesting case which resulted in an incorrect illustration that was probably a simple mistake. It is worth repeating the story, however, as an example of how complicated it can be to find and restore ancient animal remains. The story was retold in print and published in 1819 by Adams and Benkendorf.

In 1799, a huntsman from Tunguska had finished the fishing season and went to search for mammoth tusks along the coast. Although he did not investigate it at the time, he did see a frozen mass in a block of ice, so it was only a year later when the object was becoming more visible as the ice melted that it became of interest. It was only by the end of 1801 that the side of a mammoth was visible complete with two tusks. By 1803, the weight of the mammal caused it to fall free from the ice and come to rest on the sand. In 1804, the huntsman, Schmachoff, cut-off the tusks and traded them, as was normal for this sort of find as the ivory was of value. Two years later, Adams found the remnants of the mammoth, much reduced as the local population had used the corpse as dog food, as well as the wild animals of the area scavenging it. The skeleton was only missing one leg, and the head still retained some skin and hair. The whole remnant corpse was transported to St Petersburg, where it still resides. Soon afterwards Adams travelled to Takutsk where he purchased a pair of tusks, believing them to belong to the mammoth. Although Adams does not describe the details, it must be assumed that he found the fourth limb as it was incorporated into the reconstructed skeleton. This animal became quite well known and an illustration of it was hung in the Geological Gallery of the Natural History Museum in South Kensington. It was the illustration which has caused some dispute as the tusks are apparently the wrong way round, curving outwards rather than inwards. Copies of this illustration appeared quite widely in publications (Figure 4.19), including the *Guide to The Exhibition Galleries of the Department of Geology and Palaeontology in The British Museum (Natural History)* (1890). Interestingly, this same image was reproduced in the

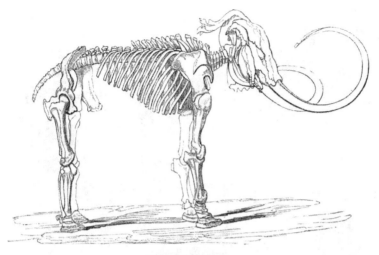

Skeleton of Mammoth.

Figure 4.19 *The illustration of the St Petersburg Mammoth with remnant skin and hair and the tusks pointing outwards. This version comes from* Vestiges of the Natural History of Creation *10th edition R.* Chambers 1853. *Source:* Chambers 1853/John Churchill/ Public domain.

Harmsworth Encyclopaedia under the entry for 'Mammoth' (Hammerton 1906), but reversed so that the animal was facing the other way to the original. There is a possible explanation for this problem with the tusk direction. In 1805, Roman Boltunov sent a letter to St Petersburg in which he drew a picture of the mammoth as he thought it would have looked in life in which the direction of the tusks were pointing outward.

These errors of interpretation and reconstruction, which seem significant with hind sight, were not necessarily so at the time they were made. One that did cause some considerable consternation for the protagonist involves a plesiosaur and a fossil hunter called Edward Drinker Cope (1840–1897). The plesiosaur was *Elasmosaurus,* and the illustrative error was one of putting the head on the wrong end (Figure 4.20). *Elasmoraurus* is similar to most of the marine reptiles of this group in having a very long neck. What Cope did was to interpret the very long neck as a long tail suitable for swimming and put the head on what was in fact the short tail. This simple mistake was pointed out to him by another fossil collector, Othniel Charles Marsh, which caused considerable ill feeling, even when a third person was called in to act as independent arbiter and agreed with March. This in itself would not have been anything but an issue of personal pride were it not that an illustration by Cope had already appeared in *The American Naturalist*, a journal in which he had a financial stake, in 1869 (Cope 1869a). This illustration was

Figure 4.20 Dryptosaurus (*formerly* Lealaps) *in tripod stance, with a short-necked* Elasmosaurus *in the foreground. Various other species are seen in the background. Source: Cope Edward Drinker (1869b)/McCalla & Stavely/Public domain.*

a pictorial representation of a group of dinosaurs, one of which was *Elasmosaurus* (probably *orientalis*) with a short neck and a very long tail.

The situation was complicated by a publication, also in 1869, which was *Synopsis of the Extant Batrachia and Reptilia of North America Part 1*. This was published in a bound cover by McCalla and Stavely of Philadelphia in 1869 as *Transactions of the American Philosophical Society* Vol XIV (Cope 1869b) and contained an illustration of the skeleton with the head on the wrong end (Figure 4.21). As this was published almost at the point of realisation by Cope of his mistake, he attempted to buy back all the published copies, although this was not entirely successful. By the time that *Transactions of the American Philosophical Society* Vol XIV, 1–252, was published in 1870, the illustration had been changed and the head was on the correct end. The figure numbers were the same and the illustration was the same size as in the incorrect version, Plate II, figure 1. This change in the paper was possible because this single article took up the entire edition. As there are still copies of the original article from McCalla and Stavely, publishers, in circulation, it is worth looking at the illustration to compare the two images.

It is easy to assume that putting the skull on the wrong end of a skeleton was a simple mistake, but if the bones were scattered, as is common with fossilised remains, the mistake might be due to it being thought of as related more to *Mosasaurus* or *Clidastes*. Both of these have tails which when extended are longer

Figure 4.21 Elasmosaurus *with the head attached to the wrong end of the skeleton, as shown in the original paper by Cope, E.D. (1869) Fossil reptiles of New Jersey. Source:* Cope (1869a)/The University of Chicago Press/Public domain.

than the neck. Their long spines extended into the tail, ending in a fluke, which at the time was not recognised in Mosasaurs, and so they were often illustrated as big eels with flippers. This assumption of a long backbone in a fossil being exactly the same as a long backbone in a modern lizard has parallels in the reconstruction of fossils of land reptiles. This is well shown in the 1854 depiction *Megalosaurus* as still displayed at Crystal Palace in London. Here is a large, strangely hybrid, quadruped and not as it is now thought to be a bipedal Therapod. There was a Victorian trend to have large dinosaurs displayed as large, lumbering, quadruped animals, in line with the known characteristics of large mammals such as rhinos and elephants. When it became apparent that both limb size and skeletal structure implied a different stance a very different animal emerged from the reconstructions.

One group that was difficult to present was the flying reptiles as there is no comparable group extant today. With no easily compared species against which to model the skeleton for form and function, it is surprising that they were accepted as flying reptiles so early after their discovery. Cosimo Collini, Director of the Elector-Palatine Museum at Mannheim, described the very first Pterosaur in 1784 which had come from deposits at Solenhofen in Bavaria. He described the skeleton as belonging to a marine reptile which used its long front limbs as paddles; this idea was still voiced on occasion up until 1843. This was sometimes suggested even after Georges Cuvier had seen the original fossil and described it as a flying reptile in 1801. Cuvier also considered that with such large eyes, they were probably nocturnal. A specimen recovered in Germany in 1809 was described as Ptero-Dactyle, which became a formal genus as *Pterodactylus*. It was not universally accepted that these were reptiles and in 1830, Newman even suggested they might be marsupial bats (Figure 4.22). The group causes some problems of origin because although the Pterosaurs were in existence through most of the Mesozoic, from the late Triassic to the end of the Cretaceous, pre-dating birds by 100 million years and bats by 150 million years, they are more closely allied to the Crocodilia than the Dinosauria.

The association of Pterosaurs with aerial life was not limited to scientific papers or comment. In 1862–1863 a story, *The Water Babies* by Charles Kingsley was serialised in a weekly magazine and finally presented in book form in 1863. Kingsley was a well-educated man and an early proponent of Charles Darwin's theory of evolution. None the less that he made reference to Pterosaurs in what was seen as

Figure 4.22 *Illustration of Pterodactyls as marsupial bats by E. Newman (1843). Source:* E. Newman.

a book for children indicates that such things were in the public mind. Although the relevant passage continues beyond the quoted section, the important part is shown below.

> Did not learned men too hold, till within the last twenty-five years, that a flying dragon was an impossible monster?

Until Cuvier had studied the fossil, it had even been conjectured that it might have been a bird. The illustrations of Pterosaurs over the next century most definitely had them as flying animals, but ideas as to their aerial ability ranged very widely. They were portrayed with a range from only being able to live on cliffs due to their huge size, to a more recent idea that even the very largest species could take off and land on a horizontal surface. If this latter point is true, then they would easily claim to outperform birds, the Great Bustard (*Otis tarda*) being the largest bird capable of making it into the air from a static position without a run at launch. It is becoming increasingly apparent that these reptiles were also quite adept on their feet, as at least partial quadrupeds. There remains a great deal of speculation over this group, both regarding their ancestry and their ability to fly, and manoeuvre on the ground. This is surprising because speculation that they were flying animals at all was quickly accepted as fact, regardless of their being no living precedent for such huge animals with strange wings. It is possible that acceptance in the popular mind was at least partly due to a continued belief, or wish to believe, in the possibility of dragons. These would be either extinct, or still out there somewhere as portrayed in Sir Arthur Conan Doyle's *The Lost World* (1912), a novel length story originally published as a serial in *Strand Magazine* before appearing in book form the same year.

Some idea of the structure and style of the wing membrane in Pterosaurs became available with the discovery of *Rhamphorhyncus* in the high-resolution lithographic stone of Germany. This certainly aided the conclusion that the wing was membranous, making it more bat-like than bird-like. The Pterosaurs quickly engendered discussion over their physiology in a way that the other extinct reptile groups did not. This centred around the simple question of whether these reptiles were warm bloodied or not. Richard Owen argued that since there were no apparent feathers, they must have been cold-bloodied. Other arguments were based on the bone structure, where there seemed to be airsacs, implying a bird-like heart and circulation, from which it was deduced that like birds, Pterosaurs were warm bloodied. It should be remembered that, at this time, the distinction between poikilothermic species and ecto- or endothermic homoeotherms was not understood and it was assumed that any warm blooded creature maintained their temperature endothermically.

With the spectacular increase in the number of fossils discovered during the twentieth century, from single celled species through the ever popular dinosaurs to birds and mammals, a much clearer insight into their structure was developed. The accumulated knowledge made it possible to describe some aspects of behaviour and, as we described with Pterosaurs, to develop ideas regarding physiology as well. While this has made it much clearer that older ideas were only partially correct, it has had some unfortunate results when it comes to depictions and popular ideas as to what fossil species, for which we have no extant analogue, were like. This has resulted in the position of speculation and guess work being taken for truth, rather than unsubstantiated supposition (Figure 4.23).

It was only with the widespread introduction of the three-colour printing process in books that it became routine to presume that colour in illustrations of fossil species, as they were thought to be in life, were true representations. This extends back to the coloured images and descriptions of the most enigmatic of all fossil groups, the Ediacaran biota. These very ancient organisms seem to represent a period before the origin of mineralised skeletons and shells and are the fauna of the Ediacaran Period, immediately prior to the Cambrian. The fossils making up this group were not recognised until the middle of the twentieth century, being soft bodied species, and therefore rarely preserved. This makes the interpretation of the fossil images very difficult for two reasons. The first is that reconstruction of a three-dimensional body from a low-resolution image is very difficult.

Figure 4.23 *A large model of a dinosaur with colouration that is an unlikely speculation, but due to lack of information cannot be completely dismissed.*

The second one is more of a public expectation. Even if the reconstruction is correct, it is expected to be given colour, even when that is inappropriate to the subject and no one can have any idea what colour these soft bodied fossil organisms would have been.

This situation extends to a certain amount through the range of fossils as they are found, but sometimes modern observation can have a significant influence in making an educated guess as to a fossil species colouring. One such example would be the pale underside and dark upper side found in sharks. Given that this is a fairly universal distribution of colour throughout the group, it would seem reasonable that it would have been found in the fossil species as well. This is not so with colours, which are so unpredictable in distribution both between species and within species, that for the most part any colour is conjecture.

Finding new fossils of already known species can make significant changes to the perception of a species and its position within palaeontology. Another way in which a shift in ideas can take place is by a very perceptive reinterpretation of already known fossils. One such change came about in the twentieth century that made a significant difference to the way we view the pre-mammalian reptiles. In 1969, John Ostrom (1928–2005) published an article that clearly demonstrated that a relatively small dinosaur, a predatory saurischian theropod in the family Dromaeosauridae, *Deinonychus antirrhopus,* was highly active and agile. This required an explanation which physiologically pointed in the direction of the animal being warm bloodied. Ostrom also revived an earlier hypothesis that at least one line of the Dinosauria gave rise to birds. Nowadays, it does seem that theropods are more likely to be the antecedents of birds that the confusingly named Ornithicians.

Within the family Dromaeosauridae, there are other predator species, such as *Velociraptor* and *Microraptor,* for which there is direct fossil evidence for the presence of feathers. These were not for any ability to fly, but for thermal insulation. In contrast to this, there is at present no indication that the single species in the genus *Deinonychus* ever had feathers. Perhaps, it is the pivotal position of *Deinonychus* in the development of a warm bloodied reptile giving rise to the avian lineage that have resulted in it being regularly feathered. Sometimes, the images are relatively muted, as in the illustration by Emily Willoughby, where the feathers are mostly brown or grey. At the other extreme is the illustration by Luis Rey where the feathers are ostrich-like and the reptile has been supplied with a birds wattles and considerable colour.

These illustrations are purely speculative, and as long as it is remembered that they are as based on reality as any science fiction illustration, there is no problem. Problems arise when it is perceived that these are partially or wholly illustrations based on known facts, which of course they are not. One area where the temptation to augment reality seems to constantly overtake the illustrator is that of colour.

That plants, especially flowering plants, and animals have had colour from the very earliest of times seems a reasonable conjecture, but to arbitrarily decide what the colours were, can cause some confusion in the mind of the viewer. Evidence of possible colour can be very circumstantial, but enough to justify inclusion of colour in illustrations. One such case is found in a paper by Bobrovsky et al. (2018) on the possibility of movement in an Ediacaran fossil, *Dickinsonia*. By extreme good fortune, fossils from the shore of the White Sea in north-western Russia seemed to have escaped the high pressure and temperatures that would normally obliterate organic molecules. Starting with the ediacaran *Beltanelliformis,* a remnant organic film implied that they were colonies of cyanobacteria. Given this association, it would be reasonable to assume that *Beltanelliformis* was of a similar colour range as extant species of cyanobacteria.

For many species, the modern approach is for colour and pattern to be introduced to skin with no reason other than for aesthetics. For some colours, there is a potential to find a true indicator of existence. The direct indicator of colour would be associated with those tones which have no pigment associated with them. These are structural colours, such as blue, that are always created in animals by diffraction gratings associated with specific physical structures. This is one of the reasons that there are so many metallic colours in animals where the light is converted into a near monochromatic wavelength. This is different to the newer finding of what appear to be melanocytes in skin impressions. This has been seen in hadrosaurs, where what looks like cells containing eumelanin have been seen (Fabri et al. 2020).

Since there is no evidence for many of the colours and patterns that appear on modern illustrations and models of species only known from fossils, it would be easy to condemn the artwork. But perhaps that would be the wrong line of reasoning. Perhaps, a better way of looking at it is that any idea without data is as valid as any other. It is certainly true that any hypothesis is a valid one to test. The problems arise when the illustrations, spotted or striped sauropods, or even simply uniformly coloured, are taken as true representations of past reality, then we have a gap between art and science. It is easy to imagine that such things do not matter, but in the future, it may be difficult to shift a myth from the public imagination. This could be necessary when new data emerge on skin tone and colour which has a direct implication for understanding behaviour of fossil species.

It is almost impossible for us to know what sort of vocalisations extinct species had. It is, however, possible for us to say what they did not sound like. Roars are most unlikely to be the usual method of communication for dinosaurs as this requires a larynx and the ability to change the sound with mouth and lips. Although some species may have had a larynx, these reptiles may have made closed-mouth vocalisations like crocodilians do. With the exception of a few species, like hadrosaurs with what appears to be a resonant crest on their heads, it is generally considered that a syrinx was used for making sound, which is the same as modern birds.

References

Adams, M. and Berkendorf, L. (1819). *Memoirs of the Imperial Academy of Sciences of St Petersburg*, vol. 5. London.

Bobrovsky, I., Hope, J., Ivantsov, A. et al. (2018). *Ancient steroids establish the Ediacaran fossil* Dickinsonia *as one of the earliest animals. Science* 361 (6408): 1246–1249.

Bourguet, L. (1742). *Traite des Petrifications*. Paris: Briasson.

Chambers, R. (1853). *Vestiges of the Natural History of Creation*, 10e. Edinburgh and London: W and R Chambers.

Cope, E.D. (1869a). Fossil reptiles of New Jersey. *The American Naturalist* 3: 84–91.

Cope, E.D. (1869b). *Synopsis of the Extant Batrachia and Reptilia of North America Part 1*. Philadelphia: McCalla and Stavely. as Transactions of the American Philosophical Society Vol XIV.

Cope, E.D. (1870). *Synopsis of the Extant Batrachia and Reptilia of North America Part 1*, Transactions of the American Philosophical Society, vol. XIV, 1–252. Philadelphia: American Philosophical Society.

Cuvier, G. (1796a). Mémoire sur les espècies d'Elephans tant vivantes que fossiles. *Magasin encyclopédique* 3: 440–445.

Cuvier, G. (1796b). Notice sur le squelette d'une très grande espèce de quadrupède inconnue jusqu'à présent, trouvé au Paraguay, et déposé au cabinet d'Histoire naturelle de Madrid. *Magasin Encyclopédique* 1: 303–310.

Fabri, M., Wiemann, J., Manucci, F., and Briggs, D. (2020, 2020). Three-dimensional soft tissue preservation revealed in the skin of a non-avian dinosaur. *Palaeontology* 63 (2): 185–193.

Figuier, L. (1862). *La Terre avanet le deluge*. Paris: Hachette. English translation, *The World Before the Deluge* (trans. H. Bristow). Chapman and Hall 1867.

Gesner, C. (1551). *Historiae Animalium* Volume 1 1551. Zurich: C. Froschauer.

Gesner, C. (1565). *De Rerum Fossilium, Lapidium et Gemmarum Maxime*. Zurich.

Hammerton, J.A. (ed.) (1906). *The Harmsworth Encyclopaedia*, vol. V, 3970. London: The Amalgamated Press.

Hutchinson, H.N. (1892). *Extinct Monsters*. New York: D. Appleton and Company.

Leibnitz, G.W. (1749). Protogaea Leipzig.

Linnaei, C. (1758). Systema Naturae Laurentii Salvii.

Mantel, G. (1825). Notice on the iguanodon, a newly discovered fossil reptile, from the sandstone of Tilgate, in Sussex. *Philosophical Transactions of the Royal Society of London*. 1825 (115): 179–186.

Newman, E. (1843). Note on the Pterodactyle tribe considered as marsupial bats. *The Zoologist* 1: 129.

Steno, N. (1667). *Canis Carchariae Dissectum Caput* (trans. Alex Garboe, 1958. London: Macmillan and Co.

Webb, W. (1872). *Buffalo Land*. San Francisco: Dewing and Co.

5

How Fossils Changed Ideas Associated with Species

In the study of evolution, the pivotal ideas are associated with Darwin's travels aboard The Beagle, studying living species around the globe (Darwin 1859). It is sometimes forgotten that investigations of fossilised remains had started people questioning the invariability of species long before Darwin started his travels. It should be obvious that to question the invariability of species requires some notion of what a species is. The kernel of this is an idea of the differences between individuals which some how separates different groups.

When thinking about species, or more precisely, early concepts of species, it is worth considering the ideas which influenced the philosophical arguments about species and biology. For this, we need to consider a primary fact that will be significant to any species concept which has its origins in the ancient world. This is that there was no background biological knowledge upon which to base an argument for what a species was. Plants and animals that were commonly met with were just there. They had been present as far as anyone was concerned forever, and the idea of disrupting a uniform picture into a series of discrete species had no clear precedent. It can be assumed from this that any early attempts in antiquity at defining species were thought experiments, more philosophical musings than attempts to structure the natural world.

We generally think that in the Western world, formal ideas regarding what a species represents started with Aristotle, and this is broadly correct in terms of documented speculation. We can be sure that private, or unwritten, speculation had taken place before the time of Aristotle (384–322 BCE). We can see some element of this in the work of Plato (427–347 BCE). Like most of the classical works contemporary with Plato and Aristotle, we usually only know them from fragments and later compilations. This is especially the case with Aristotle, where most of what we know comes from the Latin translations by Andronicus, and these were compiled and produced in the first century BCE (Falcon 2017). The influence of Plato on the philosophical output of Aristotle is not in anyway coincidental, because just as Socrates taught Plato, so Plato taught Aristotle (Figure 5.1).

Investigating Fossils: A History of Palaeontology, First Edition. Wilson J. Wall.
© 2021 John Wiley & Sons Ltd. Published 2021 by John Wiley & Sons Ltd.

Figure 5.1 *The painting of the school of Athens by Raphael Santi (1483–1520) which is a panel of the* Stanza della Segnatura, Vatican. *There are many classical philosophers represented here, but their exact representation is often open to dispute. Those which can be positively identified are the two central figures. On the left is Plato (carrying* Timaeus) *and Aristotle (carrying* Ethics). *Source:* Raphael Santi.

Much that we know of Aristotle's work regarding species stems from his philosophical ideas of defining the material world, cataloguing and interpreting what he saw around him. His ideas were broadly in philosophical agreement with the Platonic notion of forms, where forms, like ideas, are timeless and unchanging. One aspect of Plato that Aristotle did reject was the idea of a dichotomous relationship between animals, trying to use multiple characters to differentiate different animal groups. It is here that the philosophical and the practical diverge. Although modern biology would not use a concept of a dichotomous relationship between species, that is differentiating a species by a single character, we do use dichotomous keys to identify species. Plato classified living and non-living things by use of *diairesis*, which is translated in the spirit of the language as dichotomy, as this was the technique he used. Aristotle much preferred a differentiation between classes based on multiple characters, and this does give a much more robust method of creating a phylogenetic taxonomy than a simple interpretation of a single character. Much later on attempts to reconcile the ideas of Plato and Aristotle tended to result in Aristotle being seen as a single character user of *diairesis,* which was not entirely correct. Using multiple characters to determine species is one of the reasons that fossil remains do not fit into his zoology with any ease. Making the assumption that fossils are organic in origin would not have made it any easier when assessing where a fossil belonged in a taxonomic system if it, the fossil, was only partially present.

Aristotle set the scene for many later theological arguments when he declared that species were eternal and did not evolve in any way, especially one from another. While he distinguished about 500 species of birds, mammals and fish, these were deemed to have been formed as they were, which meant that to an ordered mind it should be possible to put all of the known species into a scale of perfection. In his work of the fourth century BCE, *Historia Animalium* (Aristotle 1910), there was a scale of perfection which was based on unchanging species and therefore accordingly fixed. This was an integral part of his stated investigations of what we know about the natural world, specifically animals, and therefore the causes of these distinguishing features of perfection. The application of philosophical enquiry into the animal world and the assumption that species were unchanging gave instant credibility to the idea that the natural order as, Aristotle saw it, was also unchanging.

Starting off from this premise, for Aristotle the order was simply stated; minerals were at the bottom, followed by plants, then animals and man resolutely at the top. Both Plato and Aristotle were tied up with an idea of essentialism. Essentialism refers to things being defined by their 'essence'. This is a concept which does not find a common usage in a materialist world, but one which most would recognise in an artistic context. Claude Monet produced a series of impressionist paintings in 1899 and 1900 of his Japanese bridge, *The Water Lily Pond*. The paintings of 1899 contain the essence of his vision, so although not in any way clear, it is an instantly recognisable representation of water lilies and willow trees, and he has captured their essence. An essence, therefore, is what makes a thing what it is. Being a philosophical device it is assumed by Aristotle to have a universal application; therefore by implication, rocks and mud have an essence. This would be true even if the essence remains unknown to the observer, so just as a plant or animal has an essence so does a rock. This idea, and its application to taxonomy, was certainly influential, but possibly not as all-pervading as it was at one time thought to be. For at least the second half of the twentieth century, it was assumed that the influence of Aristotle had lasted for about 2000 years in a more or less unaltered form until Darwin. This is, however, a great oversimplification and takes no account of the independent thinkers that came after Aristotle, but before Darwin.

Part of the problem with assuming Aristotle was the lynch pin of biological taxonomy is that it takes away any independence of thought of some of the greatest thinkers and naturalists of the last millennium. Aristotle features as a large persona in the story of early biology and taxonomy simply because he featured as a large figure in what we would now call a classical education, but until the nineteenth century would have just been referred to as an education. Until the late nineteenth century, for most people that went beyond simple literacy and numeracy, education would have been based upon *septem artes liberales,* the seven liberal arts. These were grammar, rhetoric, dialectics, arithmetic, geometry, music and astronomy. Other subjects were considered to be unworthy of study, or as yet just not in existence to study. It was also based upon a perpetuation of the *status quo,* which seemed to work until the industrial revolution and associated science and technology rendered the seven liberal arts inadequate for the modern world.

The long-term association between philosophy, in the form of essentialism and taxonomy, did, no doubt, have a long influence on assumptions made by theoretical speculators. These were the reconcilers of the theological world with the written accounts, often taken as true although reported at third hand. At the same time, for the practically minded, it held little to help explain the observable world. It was going to take a long time to reconcile the philosophical and the practical when it came to the species concept. Indeed, there is still a partial separation between theoretical species concepts and practical species concepts. One of the most widely used definitions of species comes from Ernst Mayr in a 1942 publication *Systematics and the origin of species from the view point of a zoologist.* His definition is

> Groups of actually or potentially interbreeding natural populations, which are reproductively isolated from other such groups. *(Mayr 1942)*

As can be readily appreciated, this definition is really only a starting point as it falls down in use with asexually reproducing species, which is implicit in the requirement of 'interbreeding natural populations'. Reproductive isolation would also pose problems because asexually reproductive groups would suggest that every clonal group would be a separate species. For palaeontologists looking at reptilian species, the definition works very poorly for practical reasons. These reasons are very easy to see when you compare a fossil with the definition by Mayr. We do not know anything about the status of the populations, or what the isolating mechanisms were for any particular fossil. More specifically, amongst modern squamate reptiles, the snakes and lizards, there are species that only reproduce through parthenogenesis. Although these are few in number, for example, only one known snake, it remains unknown how common this was in fossil species of all sorts.

The long-term influence of Aristotle did ring down the years, remaining much the same and broadly unchallenged. At the start of the Common Era, societies across much of the known world were finding their new position following the withdrawal of the Roman Empire from across Europe. Understandably, little time was given to what would have been described as natural history during this period. This was while all seats of learning in the Western world would have been exclusively for the production of clergy to perpetuate Christianity and associated social control.

With changes in financial stability of nations and the invention of moveable type for printing, knowledge became available outside traditional institutions. With increasingly wealthy nations and improving rates of literacy, time was available for education and indulgence of curiosity. In the initial years, this curiosity was exercised almost exclusively by receivers of a classical education who were nearly always also ordained clergy, such as John Ray (1627–1705) and Robert Plot (1640–1696). Sometimes the well-educated and titled, like Baron Cuvier (1769–1832), would take on the task of defining species.

It was during the time of the Renaissance, generally regarded as the fifteenth and sixteenth centuries, that there was a move from arbitrary assumptions of the existence of species to the question of whether species really exist. It was in contemporary thought, that possibly species were purely artificial, constructed by man as nothing more than a cataloguing activity. Many of the enquirers into matters of biology were Renaissance Men, both chronologically and metaphorically. This was in many ways the age of the polymath, such as Da Vinci, Galileo or Newton. Such polymaths who made up this group turned the old techniques of what was loosely science, but more accurately in modern terms natural history, into true science. They introduced a recognisable scientific method based on empirical evidence, rather than superstition. One of the most obvious and significant changes was alchemy and herbalism turning into chemistry and medicine.

This more stringent approach to investigations manifested itself in attempts to include the fossil record into biological concepts of species, simultaneously merging fossil species into a uniform taxonomy. It would be some considerable time before the fossil record was regarded as part of the phylogeny of living organisms. All through this time the nascent scientific community was still trying not to contradict the teachings of the Church, which was still a powerful influence, both socially and financially. Among the enquirers searching for order among the natural world was John Ray, who, like many of his educated contemporaries, took holy orders, in his case in 1660. In his work of 1691, *The Wisdom of God Manifest in the Works of Creation* (Figure 5.2), as in other of his works, Ray uses the term 'species' for the most definitely inanimate, such as metals, as well as for the definitely biological. Although some importance is attached to this, it was not an unusual use of the word, as it merely denotes the type of material being referred to, differentiating it from all others. So although not confined to biology, it was the definition of a biological species which Ray used that was important. This was so, because he was one of the first to try and specifically create a workable definition of a species as we would recognise it. He was not absolutely the first to try and do this as Aristotle had worked on a similar idea to define a species (Wilkins 2009). Although nearly a third of the known works of Aristotle are biological, they tend to be overlooked. We can see this as a legacy of the state of education in universities well into the twentieth century. Since subjects that were not directly regarded as part

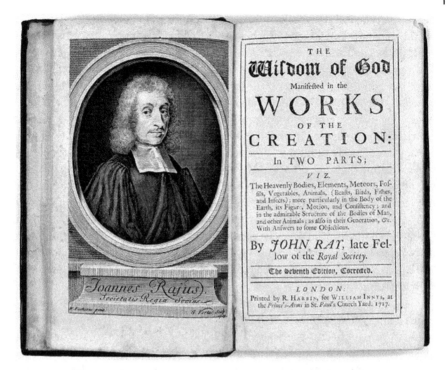

Figure 5.2 *The frontispiece of* The Wisdom of God Manifest in the Works of Creation *by John Ray. Originally published in 1691, this is from the seventh edition of 1717.*

of the classical panoply were rarely taught, it was Aristotle's philosophical output that was the basis of his fame. Neglecting his biological works resulted in a sometimes confused understanding of his ideas of species, his biological species being different to his logical species. His biological species pivots upon an ability for parents to have progeny which resemble themselves, as later picked up by Mayr (1942). This is different to his use of species in taxonomy where they constitute a group wherein variation is limited to certain characters, unnecessary to the survival of the species.

Although it has been said that between the times of Aristotle and Ray there were various observations which agreed with Aristotle and natural historians that came after him (Zachos 2016), we look to John Ray to be more precise in his definition of a species. Ray was moved to define a species using plants as his experimental material. In volume 1 of *Historia Plantarum* (1686–1704), he said that the distinguishing characters of a species are those features that perpetuate themselves via seed. This is a very biological definition as it highlights inheritance in the form of a genealogical association through generations. As a reliable and universal definition, it has severe practical drawbacks, however. To adhere to it closely as a definition, it would require the observer to witness at least two generations before a species could be declared and therefore published as a new species. While this is not generally a problem with plants as they are static and can be produced from their own seed, this is not always so easy with animals. Even so, with herd species, or the generally social, it is possible to determine associations between generations. Again, we come up against the problem of fossil species where relationships between individuals are impossible to determine with any certainty.

Ray does comment that it is reasonable using his basic form of species definition to assume that caterpillars and butterflies are different species (1713). This is even though there is an obvious association between the butterfly and caterpillar, nonetheless, they contravene part of the encompassing definition if strictly applied. It was the same definition of species which was also used by Buffon (1707–1788) in *Histoire Naturelle* published in 36 volumes in his lifetime, with more being published post-humously. This was obviously a practical, and useful definition of species as it has lasted well, notwithstanding the problems associated with fossil material.

By the time of George Cuvier (1769–1832), the complexities of creating a universal definition of species were being recognised as separate from the difficulty of identifying individual species for taxonomic purposes. Even so, the definition of species remained largely the same in Cuvier's highly influential *Animal Kingdom* (1834), where it became 'Every organism reproduces others that are similar to itself'. It can easily be seen that it is almost impossible to force fossil data into this definition of a species, but there was a way in which the problem could be resolved by self-referencing the definition. Such a process could be done in this way; if it was assumed that the various fossil remains were true species, then it was assumed that they were a freely interbreeding population of similar individuals and therefore a true species. This argument does require one of two different conditions to be true. Either, there was more than one fossil of ostensibly the same type located in the area or near vicinity of each other, and this bearing in mind that fossils are rarely complete. Or, it is assumed that the fossil remains of that particular species is a rare find because fossilisation is a rare event, and there were more of them present when they were alive.

Heinrich Georg Bronn (1800–1862) was a German biologist of considerable influence in European scientific thought in the turbulent nineteenth century. As a fervent supporter of Darwin, it was Bronn who first translated *The Origin of Species* into German. Like Lyell and Cuvier before him, Bronn was certain that species were real entities. Also like Lyell and Cuvier, when it came to higher taxons, this was not necessarily so. Very often, these higher taxons were sometimes regarded as artificial aggregations, created by obsessive taxonomy, rather than reflections of the natural order of the plant and animal kingdom.

The idea of a species being a real entity was broadly based on the idea of very limited variation within species being possible. There was an obvious corollary to this, with every species being correctly adapted to its environment, while the environment was stable, so too, would be the species and therefore of limited variability. Sometimes, the language used to describe the adaptation to a set of conditions could be misleading, for example, the use of the word 'design', which implies forethought. To think this is how it was being used would be a mistake, as at the time it was a term that reflected the biologists understanding of the 'fit' to an environment and therefore, the range of adaptions that would be expected in any given species. There was some added confusion, though, as Design was a term sometimes used as a method to justify Natural Theology. In broad terms, it was thought that whatever the creative force was, it was considered to be of sufficient power to be the cause of speciation; higher taxons were still often considered of doubtful biological value.

One of the offshoots of an acceptance of not just species, but speciation as well, was that there needed to be an equivalent acceptance of extinction. By continuation of the argument, the acceptance of extinction reflected badly on biblical literalists as it implied fallibility in God. If he was infallible, then his creations would be perfect and perfectly adapted. If they were perfectly adapted, there could be no extinction and similarly no speciation. Even without explicitly involving recognised theology, it was still assumed that there was a limit to the variation that could be tolerated before a species faced extinction. This still involved a degree of teleology, implying a final target for a species, wherein there would be limit scope for variation.

With a belief that species were real, rather than an artificial point on a continuous spectrum of variation between extremes, or a collection of similar individuals, there would be other questions that would need to be answered. One such change was that it would no longer be enough to describe a species as a freely interbreeding population. Something more was required, if it was going to be possible to define a single species out of a group of organisms. This would inevitably require the more extensive use of morphological details, such as skull, skeletal and anatomical structures. We can immediately appreciate that this would be of great benefit to palaeontologists where it is only skeletal and anatomical information that can be used to define a species. It was at this point that the definition of a species became uncoupled from methods used to describe a species.

Simultaneously with an understanding of species as true entities came the belief that there must be a gradual process of speciation. The implication of this, if it was true, was that abrupt changes in the fossil record, including abrupt changes in the general fauna, was coincidental and not a true reflection of what had actually happened. This was thought to be the case even though 'abrupt' in the case of fossil strata was on a geological scale, and therefore not so short a time as to proscribe a gradual change taking place. Such changes as there were would only seem abrupt when compared with the geology in which they were found.

It was considered that a lack of geological strata was a local phenomenon, with the absent strata accounting directly for the absence of fossils. Charles Lyell (1797–1875) also held that the fossil record was massively imperfect. By assuming this, he could account for the discontinuity of species in the fossil record, which was pivotal to ideas of a steady state of flora and fauna (Lyell 1832).

Lyell was a very perceptive scientist and used his knowledge of geology to help explain observed biological phenomena. One interesting case of this was his report on the botany of Australia, New Holland, as he refers to it, although Australia had already been used on a hand drawn map by Mathew Flinders in 1804. It was his suggestion that the plants of west and east of the continent must have a geographical block between the two sides. This was to stop the spread of plants from one side to the other, and he quotes the possibility of a great marsh, a lofty chain of mountains or a desert in the unexplored interior as the cause.

Lyell introduces complexities into the geology and biology of ancient days to explain the dwindling fossil record as you go back further in time, even though this was not regarded as entirely necessary. The reason for this is both obvious and straightforward. Even in everyday life, the farther back in time you go, the fewer artefact there are remaining, almost as though there was a natural law on the subject; so why should it not be so for the fossil record? By accepting that the fossil record was incomplete, but consistently incomplete, then it was argued that the number of species in the fossil record reflected the number of species that were extant at the time, once the losses were taken into consideration. Even with an incomplete fossil record, this also undermined the view of Jean Baptiste Lamarck (1744–1829) that change within species was gradual, as there was no evidence in the fossil record of one species changing into another. Lyell was more of the opinion that species came about through a series of extinctions and concomitant, though ill-defined, creations of new species. This was a re-expression of the observed facts, rather than an explanation of observations. As species disappeared from the fossil record, new ones appeared.

What the fossil record did show was that the apparent and very real anatomical similarities seen amongst related living species was also true of the long extinct species. This was true if even they were only known from fossilised remains and they represented hitherto unknown clades.

Although the work of Lamarck fell out of favour after the nineteenth century, he did have some observations on species and speciation which are of interest (1809). In *Tome Premier* of *Philosophie Zoologique* (1809) (Figure 5.3), he describes a species as 'every collection of similar individuals produced by other individuals like themselves'. This follows very similar lines to earlier and later definitions of a species and suffers from the same limitations of being inapplicable to asexually reproducing species. In *Tome Second*, Lamarck treads a narrow line with his descriptions of 'laws', which, of course, they are not, they should be described as hypotheses as they are untested ideas. These are separate from his laws covering evolution.

Figure 5.3 *The title page of* Philosophie Zoologique *by J. B. Lamarck. This scruffy edition of 1830 is said to have belonged to Charles Darwin.*

The first law states that every animal not fully developed can, with frequent and sustained use of an organ, develop and enlarge it. None use will eventually cause it to disappear.

The second law says that the exercised and developed organ will be passed on to succeeding generations.

Although these ideas would impinge directly upon speciation, they would only make sense in palaeontology if there was a complete and unbroken fossil record. Needless to say, they do not find favour as a method of speciation in modern biology. The ideas put forward by Lyell (1832) regarding speciation are rather different, but they also held a great deal of speculation and functional descriptions of the observed facts. Lyell was inclined to the belief that most biologists agree more with Linneaus about what a species represents than with Lamarck. In this, Lyell reported that the consensus idea was that besides being a true breeding group, a species is made up of individuals all of which have characters that distinguish them as arising from a single stock. These characters should be significant, rather than trivial, and would never be expected to vary and will remain the same from the creation of a species, until its extinction.

This idea of unchanging species was encompassed in the saltational idea of species origination. The process of saltational evolution is one which involves large steps. By implication, the small accumulated steps of gradual evolution and speciation were discounted in the process as being little more than individual variation or local variation between populations. One of the most well-known ideas of saltation is also one of the most recent and came from Richard Goldschmidt (1878–1958). His hypothesis was developed in his work of 1940, *The Material Basis of Evolution,* but the ideas that it contained were not received well. Goldschmidt was a well-regraded geneticist who introduced the concept of the intersex, and it was his work in genetics that powered his ideas.

Goldschmidt's thesis revolved around the idea that small, incremental, changes would not be enough to generate a new species. As an alternative, Goldschmidt suggested that it was macromutations, on a large scale that would effectively create a new species. This was, more or less, a concept of spontaneous speciation, which generated the phrase 'hopeful monsters'. This phrase came about because for spontaneous speciation of large, sexually reproducing organisms, large-scale saltatory jumps would need to take place twice in exactly the same way, producing one male and one female. These individuals would also need to be located geographically close enough to find each other. Hence, the phrase 'hopeful monsters'. This was not the only concept of speciation based on the sudden appearance of new forms, but it is one of the more recent ones. This was the very antithesis of classical evolution, as epitomised in a favourite phrase of Charles Darwin *Natura non facit saltum* translated as 'nature does not take leaps'.

During the nineteenth century, Lyell, like most of his contemporaries, was concerned with extinction and speciation. In his *Principles of Geology* (Lyell 1832), he quotes Buffon as saying that with regard to species 'they must die out because time fights against them' (Roger 1997). Lyell considered that the eventual extinction of species was a reasonable assumption based upon knowledge of the geology of various areas. This was most especially where volcanic activity was known and where there was a recognised fluctuation in sea level. What Lyell did was create an argument for the origination of species using known numbers of species and assumptions of rates of loss and therefore rates of creation. He makes no attempt to explain the complicated situation that would ensue from spontaneous speciation, via an individual, a process that would be tackled much later by the likes of Goldschmidt. Although Goldschmidt would disingenuously suggest that just because biologists do not think that macromutations can occur, it does not mean that they cannot. This is true, but knowledge of living forms tells us this is an unlikely situation. It would also make the palaeontologists life difficult, as a new species could only be described from a fossil find if at least two individuals could be found.

Lamarck was concerned about the tendency for biologists, more precisely taxonomists, to clump species, or split them, considering gradualism being a more likely process. He recognised that the more the 'natural objects', in the form of fossils, that were found, the more that the incomplete record of the past would be filled. This would then result in a picture of one species shading into another through time.

It was during the nineteenth century that the political and social turmoil of the times was reflected in biological thought. This sense of revolution and free thinking created a century that eventually gave way to a secular and evidence-based culture of scientific investigation. An intrinsic part of this included the acceptance of fossil material as indicators of past life and ecology which, by implication, included the generation of species. The nineteenth century saw the rise of the professional scientist and some of the earliest professional biologists, and it was difficult to reconcile their biblical beliefs with interpretations of species and extinctions, as implied by the known facts as they were at the time. One of the most interesting of these 'old guard' scientists was Louis Agassiz (1807–1873). He resisted the ideas of Darwin and believed in what we would now refer to as creationism. Even so, he did recognise the geological evidence for recent events and in 1837 he said that the Earth had seen previous ice ages. Some of his more extreme ideas, which tarnished his reputation both with his contemporaries and since then, were based around a racial divide. These came from an idea that the different races had originated separately from one another in different parts of the globe. From there, he opined that the book of Genesis only described the origins of White races as biblical scholars would only have been aware of local events, so Adam and Eve would have to have been local. It would seem that this extreme view raised some ire at the time and is now dismissed entirely.

In volume 1 of *The Evolution of Man* (1897), Ernst Haeckel (1834–1919) takes to task earlier zoologists and botanists who, as he puts it, 'occupied themselves in systematically distinguishing and describing species, without, however, any clear idea of the meaning of "species"'. This admonishment was motivated in part by a deeper understanding of biology, as it was his belief that defining a species was essential to a true idea of how plants and animals interact. Haeckel was concerned that systematic biologists were defining species as real and true, that is, pure breeding, when it was apparent that the biologist themselves had little idea of how they could be sure they were describing a true species. Haeckel also maintained that the definitions of species and subspecies, varieties and groups were merely stages in classification and of only relative importance, it was the meaning of being a species which was important. Into this argument, Agassiz was a significant voice. There were inconsistencies in the arguments put forward by Agassiz, and primarily these were associated with his ideas of the immutability of species. Much of his line of reasoning is to be found in volumes 1–5 of *The Natural History of the United States of America* (1857–1877). In the first volume, he writes his *Essay on Classification* in which he states that as Paleozoic geology shows a variety of rocks in different classes, and this must indicate that they have existed together 'from the beginning'. Following on from this, and by analogy, a consequence of having species always present surely means it can be assumed that they were not introduced successively. In his thinking on matters biological, Agassiz uses the same line of reasoning when he claims that since species are represented as whole, with no transition from one to another across epochs, then it is reasonable to state, like Cuvier, that species are fixed. At the same time in his *Essay*, Agassiz agrees with Haeckel that there is a wanton disregard of checking previous descriptions and specimens before new species are described. This almost accepts the mutability, and certainly the variability, of species.

Although the *Essay on Classification* was written two years before the publication of *On the Origin of Species by means of Natural Selection* in 1859, ideas regarding evolution were already spreading in spoken and published works. It was trying to fit observation to assumptions where Agassiz became separated from the developing mainstream of scientific ideas.

To avoid problems of species not being imutable, but changing, Agassiz thought that just as in the fossil record loss of species was sudden, so replacement species would also appear suddenly, not singly or in pairs as Goldscmidt would imply could happen, but in large numbers. This would effectively set up a complete population at a stroke in the fossil record but also underline his concept that individuals do not constitute a species but represent it. A single individual found in the fossil record, by his logic, should not, therefore, be described as a new species. All this could be summed up in his idea that 'Natural History must, in good time, become the analysis of the thoughts of the Creator of the Universe'.

Agassiz was a highly respected scientist, so he could not be lightly dismissed. The influential nature of the works of Agassiz in interpreting the fossil record in a manner which was becoming increasingly untenable, even while it was being written, created some quite harsh published commentary. Although much later, in 1897, *The Evolution of Man* by Haeckel contained the following passage which gives a flavour of his attitude to the controversy.

> It is true that in the year 1857 a celebrated and able, but very untrustworthy and dogmatic naturalist, Louis Agassiz, attempted to give an absolute signification to these categories.

The 'categories' referred to here are varieties, species, subspecies and groups. The major criticism that Haeckel had of the work of Agassiz was that instead of attempting to resolve the conundrum of classification by scientific reasoning, he leaped to religeous explanations. Or as Haeckel put it 'through the seven sided prism of theological dreams'. Haeckel seems to believe that the reasoning Agassiz was using was self-delusory, saying that Agassiz can hardly have believed his theosophic phrases. While acknowledging him as a great American who laid the foundation of much natural science and a genius, Haeckel could not imagine that

Agassiz could believe his own reasoning or has he put it 'the mystic nonsense which he preached'. As Haeckel put it:

> The divine Creator, as represented by Agassiz, is but an idealized man, a highly imaginitive architect, who is always preparing new building plans and elaborating new species. *(Haeckel 1897)*

These stern admonitions were really just a continuation of the far more detailed appraisal of the attitude to species and speciation as found in the fossil record that Haeckel had written in 1876. Haeckel, being a strong supporter of Darwin, was vehemently opposed to the philosophical ideas of past ages. These included the theological formulations that Agassiz had tried to unsuccessfully revive. Haeckel suggested that Agassiz was, in fact, the last attempt to revive the ideas of Cuvier. He did admit that it had a redeeming value in that it was the only detailed attempt to create a scientifically based teleological explanation of creation, by an eminent naturalist of the nineteenth century. Nonetheless, it was still regarded as a dubious explanation built on unsound foundations which could not be defended with the ever-increasing body of knowledge.

During this century, the nineteenth, it was generally considered that there were two, broadly opposite, ideas regarding species. The first was attributable to Linnaeus, that each species is independent of every other species, with every one of them the product of an independent act of creation. The second hypothesis, and the one which gained the ascendancy amongst most of the scientifically literate, was promulgated by many, but encapsulated by Darwin. It was assumed that there was a genetic link, a relationship between species, with the concomitant assumption that the nearer the relationship, the more they assume similarity. As there was little knowledge of genetics at the time (Wall 2016), it was more often phrased as a blood relationship, paralleling ideas of human kinship.

Haeckel was certain in his science and so he dismissed dependence upon a third-party creator as surely as he could. The vehemence of the arguments was a reflection of the new power of the rational man, no longer cowed by slavish belief in the words of the scientifically ignorant clergy. It is true that most of the naturalists of the day were classically educated, but there is a world of difference between the naturalist, who observes and makes a simple description of the observations, and a scientist who observes and makes insightful comparisons and creates testable hypotheses.

This turning point in ideas about species and speciation came with the publication in 1859 of *The Origin of Species by means of Natural Selection* by Charles Darwin. This seminal work was produced in many different editions, some of which varied considerably from the original publication. Although it is considered to be an exposition of evolution as we understand it, the species concept and what species mean in reality is dealt with in some depth. The gestation of *Origins* is interesting in itself since it could have appeared before 1859, had there not been residual issues associated with the perception of the content as anti-Church. Darwin himself claimed that the idea of species and evolution as a theoretical construct, for no hypothesis was going to be possible with this idea, was formulated in 1839. But as we now know, for various reasons, it was delayed until 1859 before it was published. Many ideas have been put forward for the delay, most of which revolve around the possibility of offending the clergy, who still had a very great influence on day-to-day life. This is probably an over exaggeration. Certainly, it would have been foreseen that the book would cause controversy. However, during the long gap between conception and publication, Darwin seems to have promulgated his ideas quite freely at meetings and in letters, without any problems of theological censure.

One of the major motivators for Darwin to come to publication was the completion of the work by Alfred Russel Wallace (1823–1913), who saw something in the fossil record which, to him, formed a pattern that needed an explanation. What Wallace described was formations of fossils which formed a pattern of distribution of different species that would be best understood by assuming that new species arose in proximity to a closely related or similar species. The problem of defining species from skeletal remains would always result in assumptions of kinship, even if they were examples of convergent evolution. Wallace (Figure 5.4) was a well-travelled biologist and his interest in species and speciation can be guaged from a letter written to a fellow biologist, Bates, in 1847. The original is archived and is quoted in Raby (2002). Wallace wrote

> I should like to take some one family to study thoroughly, principally with a view to the theory of the origin of species. By that means I am strongly of opinion that some definite results might be arrived at.

The family he refers to here is a family of beetles (Order Coleoptera), a group which even in the nineteenth century was recognised as astonishingly versatile and numerous.

When the first edition of the eagerly awaited *Origins* was published, it was quickly oversubscribed, even at the initial price of 15 shillings (pre-decimalisation there were 20 shillings in a pound or 240 pence in a pound). During his lifetime, there were several editions published, each one corrected and revised by Darwin himself. The alterations included additional material added specifically to rebuff particular points and arguments that had been raised against the central thesis regarding the process of evolution. Altogether, there were six editions of *Origins* produced in Darwin's lifetime. It was in the fifth edition, published in 1869, that the phrase, so well known in the popular imagination, 'survival of the fittest' became the aphorism which epitomised not just the concept of evolution, but Darwin himself. This was not an original term as it came from *Principles of Biology* by Herbert Spencer (1820–1903), which was published in 1864. The work by

Figure 5.4   Alfred Russel Wallace, taken about 1895. *Source:* London Stereoscopic and Photographic Company, https://id.wikipedia.org/wiki/Berkas:Alfred-Russel-Wallace-c1895.jpg#/media/Berkas:Alfred-Russel-Wallace-c1895.jpg, Public Domain.

Spencer dealt with the concept of evolution, but was primarily Lamarckian in nature, even though he generated the celebrated phrase. In the sixth edition of *Origins,* published in 1872, Darwin first used the term 'evolution' in the modern form. Until that time, evolution was primarily an embryological term. By the time of the sixth edition, sales were both good and consistent, so as a calculated means of keeping sales buoyant, the font was reduced, thereby reducing the size of the volume and the price was halved to 7/6, 7 shillings and 6 pence. The knock-on effect of *Origins* was that many other non-fiction books found an appreciative audience. But even with this buoyant interest in all the things scientific and high sales rates, it would perhaps be over-stating the case that natural history books outsold fiction in the middle of the nineteenth century, as it has sometimes been suggested (Francis 2004). Although theological criticism was significant, this was often put in terms of arguments needing rebuttal, which Darwin managed with each edition. What was more difficult to account for was the bleak outlook and aggressive survivalism that seemed to be contained in the story, describing an almost Malthusian doom for humanity. The style of writing is not always easy to read, the sentence structure not being to modern taste.

The fossil record was of considerable interest to Darwin and so he says in Chapter 6 that since geological stratum were not full of graduated intermediate links, in a fine organic chain, this is a significant objection to his theory. His original rebuttal of this suggestion was that this may be a reflection of the imperfect nature of the fossil record. Although this would be a robust argument that would be quite acceptable, Darwin seems to want to make a statement that is testable. He then goes on to suggest that it may well be something to do with what sort of intermediate fossils are being looked for. Even though, by his own admission, he tends to consider intermediate forms being phenotypically directly between two species, this, he says, is totally false. In the fossil record, what should be looked for is an intermediate form between each species and a 'common but unknown progenitor'.

This idea of looking in the fossil record for such linking species, not just as intermediates, but as progenitors of a clade, is both perceptive, and at the same time very difficult to bring to fruition. There are many practical reasons for this problem. The major statistical problem associated with this is that not finding the evidence could reflect a lack of technique or the wrong geography, rather than indicating that the fossils do not exist. To get around this, Darwin uses an analogy, citing an example from pigeon breeding to clarify his argument. Fantail and pouter pigeons, which are both descended from the rock pigeon (*Columba livia*), were the two extreme ends of his physical range. So if we had every intermediate between the two fancy pigeons that had ever existed, we would have an extremely close series among fantail pigeon, pouter pigeon and the rock pigeon, but no variety directly intermediate between fantail and pouter pigeon. Such a hypothetical intermediate form would have a partial fantail (fantail pigeon trait) and partially inflatable crop (pouter pigeon trait) in the same bird. This example is a clear demonstration of Darwin's thinking, and because we know where these fancy pigeon varieties originated we know what an intermediate would look like. Neither of these situations can be clearly stated for animals described from the fossil record where the entire clade is extinct.

The second example Darwin gives is between the horse and the tapir. In this case, he states that there is no reason to believe there was ever an intermediate between these two species. He does, however, state that there would be a common ancestor which may have differed considerably from both of them. These parent species are generally extinct and, so too, are their preceding species. Although going back far enough into the Perissodactyla does bring us to the extinct Palaeotheridae and possibly the genus *Hyalotherium*, the link between *Equus* and *Tapirus* via the brontotheres makes this particular example difficult to visualise. It may be tricky to see the details held in a fossil record, but this does serve to underline a comment that Darwin makes in Chapter 6 of *Origins*. He comments that anyone who has read *Principles of Geology* by Charles Lyell and still does not admit to the incomprehensibly vast time that must have passed 'may as well close this volume'. Darwin gives many examples of deficient fossil series and possible explanations for this. One very good point which he makes is that the area of the globe so far examined for fossilised material is relatively small. In this, it is certainly true, as palaeontology has become a more widespread and popular science, so the number of specimens and species that have been found and are being found has continued to grow.

With the description of extinct species known only from partial remains, the difficulty of determining what constitutes a species amongst the disappeared is not easily resolved. With no guidance on the status of a specimen, other than the extant skeletal remains, with which to determine the taxonomic position, whether species or not, becomes a very difficult task, fraught with confusion. The duality of the philosophical and the pragmatic remains a constant palaeontological conundrum. This has to be the case if there is any question of a species being nominally defined as a group isolated by reproduction, because with that definition there is no way of knowing which extinct individual belonged to which species. This is not a problem confined to extinct species but is also the case with very rare species, where it may not be possible to be certain whether physically separate populations represent different populations or different species. The giant salamander from China, *Andrias,* the worlds largest amphibian, was once considered to be made up of two species (Webb et al. 1981), but through genetic analysis it is now thought to be made up three or more species. The dependence on genetic analysis may not be entirely sound as these allopatric species are capable of hybridisation, which may imply they are not entirely differentiated from their progenitor species. The clarity assumed by genetic analysis may not in itself be adequate, so for extinct species where such detailed analysis is not possible, the problems of taxonomy will probably always remain.

Defining a fossil species from morphology alone is always going to be problematic, because like the salamander, certainty is going to be just out of reach. For some fossil forms, which correspond to groups that are still extant, such as insects, high-resolution wing fossils can lend some certainty to the taxonomy. Broadly, the situation for the palaeontologist transcends how species are defined and moves into how a species is identified, a practice which for most species, whether extinct or living, is usually based on physical similarities (Richards 2010). The modern techniques of DNA analysis, which are much favoured by the modern 'splitters', are, of course, useless in palaeontology. One thing that remains essential to palaeontology is that species as defined by morphology represents an association between groups that pre-date the fossil artefact being studied. In this, it becomes part and parcel of our clarity of understanding of evolution.

References

Agassiz, L. (1857–1877). *The Natural History of the United States of America*, vol. 5. Boston: Little, Brown and Co.

Aristotle (4th Century BCE) (1910). *Historia Animalium* (trans. D'Arcy Wentworth Thompson). Reprint Gianluca Press (2020).

Cuvier, G. (1834). *Animal Kingdom* (trans. H. McMurtrie). London: Orr and Smith.

Darwin, C. (1859). *On the Origin of Species by Means of Natural Selection*. London: John Murray. Many modern reprints available.

Falcon, A. (2017). Commentators on Aristotle. In: *The Stanford Encyclopedia of Philosophy* (Fall 2017 Edition) (ed. E.N. Zalta). https://plato.stanford.edu/archives/fall2017/entries/aristotle-commentators.

Goldschmidt, R. (1940). *The Material Basis of Evolution*. USA: Yale University Press.

Haeckel, E. (1876). *The History of Creation*, vol. 2 (trans. E. Ray Lankester. London: Henry King and Co.

Haeckel, E. (1897). *The Evolution of Man*, vol. 2 (trans. E. Ray Lankester. New York, USA: D. Appleton and Company.

Lamarck, J.B. (1809). *Philosophie Zoologique*, vol. 2. Paris: Dentu.

Levine, G. (2004). Introduction. In: *The Origin of Species* (eds. C. Darwin and G. Levine). New York: Barnes and Noble Books.

Lyell, C. (1832). *Principles of Geology*, vol. 2 (trans. J. Murray London. Reprinted University Chicago Press 1991.

Mayr, E. (1942). *Systematics and the Origin of Species from the View Point of a Zoologist*. USA: Harvard University Press.

Raby, P. (2002). *Alfred Russel Wallace: A Life*. USA: Princeton University Press.

Ray, J. (1686–1704). *Historia Plantarum*. In three volumes. St Paul's, London: The Printers of the Royal Society.

Ray, J. (1691). *The Wisdom of God Manifest in the Works of Creation*. St Paul's, London: W. Innys.

Ray, J. (1713). *Three Physico-Theological Discourses*. St Paul's, London: W. Innys.

Richards, R. (2010). *The Species Problem: A Philosophical Analysis*. Cambridge University Press.

Roger, J. (1997). *Buffon: A Life in Natural History* (trans. Sarah Bonnefoi). USA: Cornell University Press.

Spencer, H. (1864). *Principles of Biology*, vol. 2. London and Edinburgh: Williams and Norgate.

Wall, W.J. (2016). *The Search for Human Chromosomes: A History of Discovery*. Switzerland: Springer.

Webb, J., Wallwork, J., and Elgood, J. (1981). *Guide to Living Amphibians*. London: The Macmillan Press Limited.

Wilkins, J. (2009). *Species: A History of the Idea*. California, USA: University of California Press.

Zachos, F.E. (2016). *Species Concepts in Biology*. Switzerland: Springer International AG.

6

Fossils and Evolution

The idea of evolution in the way we know it now is a very recent philosophical construction, but it has roots going back much further than is usually credited. With older uses of the word evolution, care should be exercised, as very often the meaning is different to its modern usage. There is no doubt that evolution became a complete theory in the mind of the public with the publication of *On the Origin of Species* by Charles Darwin in 1859 (Darwin 1859). One of the questions associated with evolution as a whole would be addressed later, when humanity became recognised as part of the same process, if not yet as part of the biosphere. The question, of course, is where we, as humans, fit into the natural order? It had been a long-running question, often stepped round in deference to religious sensitivities. When finally asked, the answers were many and various, with humanity generally being placed outside the general order of nature; from our own self-assessment, we were superior in every way and therefore overlords of the world. As we recognise now, such hubris brings major problems for every species, including ourselves.

Although evolution traditionally deals with the development of species and clades, with their proliferation and extinction, the origins of life were rarely looked at. This was probably because this is an area which was at the time, and still is, an entirely unknown quantity. Even so it is worth looking at early ideas of the origins of life to see what was inherited by religious groups and turned into their myths and dogmas.

In the first years of a structured civilisation, where time became available for pondering what was essentially philosophical questions, one in particular seems to recur, 'where did life come from'? This was a question that was going to be addressed, but inevitably without an answer. It would be a subject of speculation for millennia, as there were no pointers as to where to start in finding an answer. Nonetheless, speculation was always possible since no answer could be ruled out using facts alone. A very early suggestion for the origin of life came from Anaximander in

Investigating Fossils: A History of Palaeontology, First Edition. Wilson J. Wall.
© 2021 John Wiley & Sons Ltd. Published 2021 by John Wiley & Sons Ltd.

about 520 BCE. It is sometimes suggested that his ideas pre-empted the ideas of the nineteenth century evolutionists. His conceptual framework was, however, both simpler and more pragmatic, as it tried to explain the observed world in the simplest way possible, while still answering that very first question of where life started. Anaximander formulated the idea that living organisms were generated from what we would now term primaeval slime (he would not have referred to it as Primaeval, as being from Iona and taught by Thales he would have spoken Greek and Primaeval is of a Latin root) and the heat of the sun. According to the same scheme of thought, later species emerged out of prickly husks onto dry land.

Although this may somehow just about imply evolution from one species to another, this was more descriptive of serial creation, culminating in Man. This may have been a philosophical construction to answer a question, but it had no intrinsic mechanism of inheritance and no idea of consistent species. With so little data, except for the snapshot of life that they lived with, giving an impression of permanence, Anaximander's hypothesis for the origin of life could only be coincidental with modern thought. The only thing that was known was that all things break down when dead; it was not even clear to these ancient observers that species also bred true.

After Anaximander, a more detailed set of ideas was put forward by Leuccipus and Democritus for the origin and perpetuation of life. This was still recognisably philosophical speculation, rather than a scientific investigation as we would recognise it. Their school of philosophy is sometimes considered to be the origin of the atomic theory of matter; again, this is rather unlikely as the range of things which could be seen was limited to the naked eye. Nonetheless, it is easily possible to see progressively smaller particles of soil or sand, until they are no longer visible, but can be felt as rough material, or seen when in large enough amounts, such as is found in clay. This set of ideas started from a premise that nothing arises from nothing and consequently cannot be reduced to nothing. Similarly as the void is infinite, so, too, must be the number of atoms, which come in all shapes and sizes, but all of which are too small to be seen. In this philosophy, all things have to be explained in terms of their atoms. In this way, and consistent with philosophies having no experimental background, it explained all things in a way which could not be tested. Probably, one of the most important aspects of this world view was the idea that some things were simultaneously unseen but not supernatural. When Aristotle was working, he was primarily an observer and as such denied the easy appeal of the atomists, Leuccipus and Democritus, he expected to be able to observe all things and not take them on trust, like particles disappearing into the infinitely small.

What Aristotle did demonstrate was an understanding of the interactions between species – a sense of what would become ecology. To Aristotle, every species was complete, which would be quite reasonable since no change was perceptible in the plants and animals around him. The logical completion of this idea was that he could not have any concept of evolution; if animal groups do not change, they cannot evolve. Even so, for Aristotle, all species could be arranged from the simplest, or lowest, form, to the highest, or most complex. The implied relationships between species contained a considerable teleological content, which often still turns up in modern conversations about evolution. This can be seen in phrases such as 'adapt to a changed environment' implying a change which can be directed towards a goal. This is a teleological statement as there are no directed changes, only random ones some of which are advantageous. What Aristotle did approach was another question that had puzzled many thinkers and philosophers; why organisms gave rise to progeny of the same form and never of a different sort. This was a question that would run onwards for centuries and since there was no such thing as palaeontology by name, all of the investigations of fossils and extant species were carried out by naturalists. The easiest explanation was therefore that the natural arrangement of species was Divine, an escape from critical thinking which avoided clerical sanction at the same time.

This position regarding evolution, that it does not exist, would remain the *status quo* for many centuries in biological thought. As Haeckel put it with regard to Aristotle in 1897, *During the two thousand years after Aristotle no essential progress in zoology in general, or in the History of Evolution in particular, is to be recorded.* This is not really so surprising as the period covers many centuries of social and political turmoil, when intellectual curiosity took second place to survival in a society where power and wealth were increasingly concentrated in the hands of progressively smaller numbers of individuals. It was during this period that the works of Aristotle were copied, translated, sometimes even added to, but not significantly improved with any original investigation. There is no doubt that throughout this period, scientific research of any sort was clearly defined by the dogmatic nature of the Church. This extended from the geocentric nature of the solar system to the bull of Pope Boniface VIII (c. 1230–1303, Pope from 1294 to 1303) which threatened excommunication for anyone guilty of dismembering a human corpse. It was only with the Reformation that it became possible to break the dogma of the Church. Even so, this was done carefully and in very small steps so as not to tread too heavily on toes that might officially sanction investigation, but may well still resent the implications. It is sometimes said that the last great apologist for divine creation was Louis Agassiz, who was also one of the last holders of constancy of species.

It was in the nature of the question, and the impossibility of perceiving the immense time scales involved that the difficulty of assimilating a comprehensive and understandable theory of evolution manifested itself. It is frequently said that Darwin had many precursors who either had all the relevant data and did nothing with it, or nearly had sufficient information to come to the same conclusions as Darwin, but took a wrong turn (Eiseley 1958). This is, however, putting a gloss on

Darwin's predecessors work which for the most part it does not deserve. One of the most influential of these Darwin detractors was Samuel Butler (1835–1902) who championed pre-Darwinian thinkers. Although it is easy to criticise alternative ideas, in the case of Butler, it did cause considerable problems for contemporary scientists, considerably out of proportion to its scientific merit.

The reason that it had such exposure and accidental influence was that Butler (1835–1902) was a well-known and accepted literary figure who had studied classics at St Johns, Cambridge. So although scientifically uneducated, he had a profile which ensured that people at least considered his musings. The most significant of his works was *Erewhon* (Butler 1872), a satirical Utopian novel. The title was supposed to be 'nowhere' backwards, notwithstanding the reversed position of the w and h. This was published anonymously in 1872, and this and other works gave him a literary gravitas which he then lent on and extended into a scientific commentary. His ideas, as expressed in *Evolution Old and New* (Butler 1979), on evolution antagonised both the Church and the scientific community. Broadly, Butler claimed that Darwin had borrowed from previous authors and scientists. The specific individuals that Butler was championing were Erasmus Darwin, the grandfather of Charles, Buffon and Lamarck. The criticism was mainly motivated by Butler being against natural selection, rather than evolution as a concept. His understanding seemed to have stalled with Lamarckism, that easy to understand and comprehensible explanation without proof. We can see clearly that the criticism by Butler stemmed from his scientific ignorance; it was this transparent lack of understanding of the underlying principles of evolution which resulted in his ideas not being taken seriously in scientific circles, but were regarded as popular entertainment, an idea that Butler himself agreed with. His book *Evolution Old and New* was reviewed by Alfred Russel Wallace in the year of its publication (1879). This was not a particularly encouraging review, but the book was still widely read. It would be a mistake to think that this disparaging of Darwin's work, by trying to find predecessors whose work predated Darwin, did not stop with Butler, it carried on into the twentieth century (Eisley 1958). Many of these critiques of Darwin were based on taking previous work out of context or putting sections together without intervening parts. In this way, it was possible to endow articles with meanings and implications they never originally contained. As Bowler points out (1983), many of the pre-Darwinian natural historians were from an age when it was culturally impossible for them to consider a process of speciation which did not include a deity. Similarly, any idea of evolution would have been extremely difficult to consider. Much more likely, religious niceties would have rendered such a question so powerless that it would not even have been thought about.

The content of works by Buffon, though perceptive in observational terms, was held in check from taking the intellectual leap of making definitive conclusions regarding the changes in the planets flora and fauna over time. It was the inability of him to extrapolate from observation to hypothesis which marks them out as not being a particular influence on Darwin. Indeed, Darwin did say that he was not familiar with the works of Buffon. By the time of the fourth edition of *Origins*, he referenced Buffon as being an early user of evolutionary thought. Buffon made these evolutionary comments by discussion, but without any explanation or suggestions of a mechanism by which they could occur. He was certainly the first to bring questions of what happens during large periods of time with respect to the world and associated species, to a wide scientific audience (Buffon 1778). Buffon also wrote of 'struggle for existence' and later 'struggle for survival', these being terms that were in use during the late seventeenth century. Buffon seemed to believe in a form of Lamarckism. He proposed that since the inception of a species, they could either improve or degenerate on dispersal from the point of origin of the species. This is why he suggested that all quadrupeds originated from a very small original group (Roger 1989).

The creationism that Buffon encompassed in the eighteenth century certainly put the brakes on independent secular thought until well into the nineteenth century. It was the burgeoning development of the industrial revolution and the belief in the infallibility of man which fuelled a belief in the ability to understand and control nature. Part of this determination to understand the natural world without recourse to a deity resulted in some quite strongly worded reproaches towards works which included a religious component. Thus, it was that Haeckel (1897) referred to Louis Agassiz (1807–1873) as '..gifted with too much genius actually to believe in the truth of the mystic nonsense which he preached'. Before this, Haeckel had said that with *Recherches sur les Poissons Fossiles* (Fossil Fish), Agassiz had given great impetus to the still young science of palaeontology. This was true of many of the collectors and explainers of fossils of the early nineteenth Century. Their collections and descriptions of fossils were superb, but their moral sensibilities engendered by a religious education caused problems for them. They were keen to try and square the circle with ever more bizarre attempts to side step ideas of mutable species and extinction. This manifested itself as complicated explanations, always ending with the guiding hand of a deity. This seemed to satisfy the writers, but ignored the basic premise that if a grand Creator was involved, no explanation would be necessary because what is being described is magic by any other name, and magic needs neither scientific explanation nor justification.

Contradictions are apparent within the works of Agassiz which put him in a position diametrically opposed to the work of Darwin, which considering his scientific work, is a surprise. There is no doubt that along with Cuvier, the early work of Agassiz was of immense importance to the developing science of palaeontology (Agassiz 1833-1843). More than that, it also increased in the mind of the public and made acceptable, the whole idea of a great age for the Earth. He also developed the idea of the existence of a past fauna now lost, but still recognisable from the fossil record. One of the aspects of *Origins* which is sometimes forgotten is that with each new edition, Darwin made changes which would address specific criticisms of the previous edition. In this way, he would try to create a new line of argument that would counter the ideas of the hard to convince. Such reviews and revisions of the text did not make specific reference to the critical individuals. The only time that Darwin wrote a public admonition regarding a specific criticism was in 1880 (Darwin 1880). In this paper in *Nature,* Darwin takes exception to a phrase that Charles Wyville Thomson writes in *The Voyage of the Challenger* (1877). Thomson was the scientific supervisor of the expedition which took its name from HMS Challenger. The phrase which moved Darwin to a public demonstration of surprise and annoyance was

> The character of the abyssal fauna refuses to give the least support to the theory which refers evolution of species to extreme variation guided only by natural selection. *(Thomson 1877)*

Darwin's response was clear. He says this is a common criticism reached by theologians. He asks, rhetorically, if Sir Wyville can name anyone who says evolution depends only on natural selection? It was the use of *extreme* variation which made Darwin uncomfortable, citing both his book and the lack of extreme variation in domestic species.

Carles Lyell, who was predominately known as a geologist, came into a system of biological thought where the ideas of Cuvier had held sway for a long time. It was Cuvier who dominated biological thinking more or less until the advent of Darwin's work in 1859 (Haeckel 1876). This seems to be ascribed to a greater, or lesser, extent on the sheer power and dominance of Cuvier and his legacy. There was an increasing belief that the overall story which Cuvier espoused was flawed. They were drawn from an unsupported assumption that during the entire history of the world, there had been a series of catastrophic changes, of indeterminate type and cause. During each spasmodic event, more or less all of the currently extant animal and plant species would be annihilated. After each event, a new set of animals and plants was said to have been created, which of necessity to the theological nature of the event, would have to be more advanced and better than the last round of creation. Cuvier's catastrophism could not be maintained against the gradually accumulating data, mostly from zoology and palaeontology, which etched away at the central dogma of Cuvier's idea of step-wise evolution, and yet it carried on as a primary explanation for a generation.

When Lyell published *Principles of Geology* in 1832, he was setting a scene that was going to be a considerable influence on the train of thought that would lead Darwin to his own conclusions. What is sometimes lost is the acceptance, not just retrospectively, but by his contemporaries as well, that Cuvier was not a tenable explainer of the progress of the earth through geological time. This may seem in strange contradiction with the idea that Cuvier reigned as the dominant thought in biological teaching for the first half of the nineteenth century. However, it takes time, even in the twenty-first century for radical ideas to move into mainstream taught courses. Besides, while Cuvier was altering perceptions of the process of speciation and biological complexity, Lyell was essentially altering the way in which geology and the age of the Earth were being thought about.

This is not to assert that everything the modern men of the nineteenth century said or propounded was correct, or unhindered by the cultural mores of the time. Lyell was keen on the idea of a uniformitarian world that was in a steady state of being, broadly unchanging. This does not rule out fluctuations in temperature, even what we would think of as extreme fluctuations, which Lyell was aware of having happened in the distant geological past. Such variations as did happen were associated, according to Lyell, with two major processes. The first was the distribution of seas reflecting sunlight. The second was land masses absorbing the solar heat. Although this was his primary opinion of temperature changes during the Earths long existence, it also affected his ideas on the biology of the planet. This idea is sometimes referred to by the encompassing, but cumbersome term, *nonprogressionism*. The fundamental argument that this propounds is simply that any movement of life forward, as though progressing in a specific direction, say towards greater complexity, was unsatisfactory. If nonprogressionism was real, by implication, there would, therefore, have to be mammalian fossils to be found in amongst those of the great ages of the dinosaurs. There was a suggestion of everything being present at the same time.

Such fossils as would reinforce this hypothesis were claimed to have been found from what we would now describe as an inappropriate time, although these were later demonstrated to be incorrect identifications. What was more interesting about the argument that Lyell put forward for not seeing mammalian fossils was the two-pronged approach he used in his reasoning. Primarily, the argument was made, quite correctly, that not all organisms left fossilised remains and as such, it was quite possible that this is why no mammals had so far been found. As an argument, this has significant holes in it, but superficially seems reasonable. The other suggestion Lyell made was what would now be seen as a form of the null hypothesis in that just because they have not been seen, it does not mean they do not exist. This particular hypothesis can be disproved by a discovery, but in statistical terms, it becomes a probabilistic exercise. Put another way, the more you search without a positive find, the more unlikely it becomes that such a discovery will be made. The probability never becomes zero, but approaches it ever more closely. This statistical approach to his argument would have been one which Lyell would have been unaware of as statistical analysis was at a very early stage of development. These two arguments could be thought of as expressions of the point that the fossil record is both imperfect and incomplete, both significant points for the palaeontologist.

The lack of fossil evidence was incorrectly considered by Lyell to be a tacit agreement with his idea of nonprogression in evolutionary terms. This came under considerable stress as a workable hypothesis when Darwin had published his work on the origins of species. At this time, evolution was still very much a description of what was being observed in the fossil record, constructing a functional explanation to cover the known facts. What was distinctly lacking was a mechanism which could explain the process. Such a mechanism, which we now refer to as genetics, was first described in 1866 by Gregor Mendel, although the significance was largely missed until the turn of the twentieth century. It was then that a translation was published (Bateson 1901) which launched the whole idea of inheritance onto a new trajectory. This new science was not immediately of significance to evolutionary thought, but it was significant in understanding how variation within a species could occur. It should be remembered that although a mechanism had been described by Mendel (1866), it would not be until almost the middle of the twentieth century that it would be demonstrated what the chemical messenger was for the process of inheritance. So, it was still in part a functional description, to many it was a metaphor, rather than a hard reality, and it remained obscure exactly how it could operate above the level of individuals or species and influence evolution.

With knowledge of inheritance in the state that it was during the nineteenth century, that is virtually nothing other than a nebulous notion of an organism giving rise to organisms of the same kind, it would be logical for two major proponents to coexist. In some cases, Lamarck's and Darwin's ideas could even become intertwined. One of the biggest ideas of the late eighteenth and early nineteenth century was Lamarckism; in evolution, this had two broad rules put forward in *Philosophie Zoologique* (1809). The first was that organs and tissues were strengthened and developed by usage, and the second was that acquisition or losses caused by either use or disuse are passed onto future generations. Without any knowledge of mechanisms for control of inheritable traits, this is not an unreasonable pair of assumptions to make. It is this lack of understanding which can quickly change conversations of evolution to conversations of Lamarckism with no perceptible faltering of step.

That Darwin could see beyond these two simple concepts put forward by Lamarck, while nodding to their plausibility was a significant reflection of his deep biological knowledge (Figure 6.1). Along with these two tenets, Lamarck also had an element of spontaneous generation at the lowest phylogenetic level, as an attempt to explain how there was always what was regarded as small and lowly life. This was necessary to fulfil his idea of a ladder of complexity along which organisms moved in an upwards direction. Lamarck was not the first to

Figure 6.1 *This edition of* Philosophie Zoologique *was published in 1850 and is significant in being Darwin's personal copy.*

suggest an idea of evolution in biology, but he was the first to develop this from a simple idea to a complete theory. Given the limits of knowledge at the time, it was a useful exercise that could not be gainsaid until more information was available, so it survived almost unchanged for many decades. There was some support for Lamarckism from Lyell as he favoured long time scales in his view of the world, which would allow organisms to change by usage.

All of the ideas surrounding evolution which were formulated in the nineteenth century had some credibility, but also, either consciously or unconsciously, a considerable teleological content. Part of the problem was that, as Lyell asserted, we could never see the original state of the world. He also asserted that only observable causes could affect the world, assuming anything else would be unscientific. As only slow processes were observed, then only slow changes, building structures slowly, were assumed by him to be the truth of geological activity. In this, there could, therefore, be no sudden breaks or changes in the observed continuity as this is outside our normal experience. This was a line of reasoning that Darwin also employed, only processes which can be observed in extant populations and groups can be projected backwards through time to explain both speciation and evolution. A clear idea was forming that what could be seen happening now would have been happening in the past, results happening now would also be happening in the past.

Lyell poses a question regarding the variety of species which is interesting by its assumption of collaborative change within a species or population. He set a question to the effect that if all the higher orders of plants and animals were comparatively modern, derived from more simple organisms, why were there still any simple organisms? It was obvious to Lyell that a further hypothesis was required to explain why there were still so many of the lower forms still in the world (Lyell 1832). The answer given, by Lyell, in response to this question is a convoluted one requiring a teleological account of nature being obliged to continuously produce organisms of the lowest kind, to which complexity and additional organs can be added during the slow passage of time. This was essentially a Lamarckian view of biological change, with the origination of life still ignored as a process. Lyell also had a problem with the idea of man being intrinsically linked to the animal world through an evolutionary progression. His ideas of the animal world being in a steady state required the assumption of humanity being in some way a separate creation to plants and animals, and also by implication, recent. The question of how mankind fits into the phylogeny of life is one that was going to cause argument and debate for many years. A geologically recent origin for humanity was assumed by virtue of there being no fossil remains found which could be linked to a primate phylogeny with human-like characters. This was a situation that was not going to be altered until 1936 when the first hominid fossil was discovered, in Sterkfontein Valley in South Africa. The discoverer was Robert Broom (1866–1951), and the remains he found were described as *Australopithecus africanus*.

While animal fossils were well-known and captured the imagination of the public throughout the nineteenth century, the study of palaeobotany was also accumulating data which added to our understanding of life and evolution. In 1804, Ernst von Schlotheim (1764–1832) was one of the first to publish illustrations of carboniferous plants, but it was nearly 20 years later that Adolphe Brongniart (1801–1876) made palaeobotany as important as palaeozoology. Brongniart had developed a detailed knowledge of modern botany, especially of flowering plants, long before he turned his attention to fossil plants. Adolph was the son of Alexandre Brongniart, a colleague of Cuvier and of some repute in his own right. Adolphe Brongnart established his credentials as a palaeontologist with the publication in 1822 on the distribution of fossil plants (1822). Later publications cemented his reputation as a considerable authority on this subject. What Brongniart concluded was that there had been four botanical periods in the history of the world, with discontinuities between each one. Just as after the great age of reptiles, we retain a considerable reptile fauna so, too, after each botanical period, there was retention of representatives left from each period. It is of interest to recap the different botanical periods, as their construction has relevance to the developing picture of species as evolving over time.

The first period, broadly the upper Palaeozoic, is the time in which the dominant vegetation was made up of vascular cryptograms. These are members of the Pteridophyta, vascular plants that reproduce by spores. The second of Brongniart's botanical periods was a period of impoverished flora and the appearance of the first conifers. By the third period, which was approximately the Mesozoic, there appeared the first cycads, while the bulk of the flora was made up of conifers. By the time of the Cenozoic, Brongniart's fourth period, not only had the flowering plants appeared, but they had also started to dominate the botanical picture. This trail of perceived advancement in the flora was a pleasing outcome for Brongniart who saw it as paralleling the same progressive sequence that had been developing in the study of animal fossils.

Trying to ease the sensibilities of nineteenth century religious feelings and conflicts was always going to be difficult. There was, however, one individual who took up the challenge. This was Robert Chambers (1802–1871). He was the brother of William Chambers, with whom he started W and R Chambers, Publishers. In 1844, Robert Chambers published *Vestiges of the Natural History of Creation* (1844), great lengths were taken to preserve the anonymity of Robert and it is said that only four people new the true identity of the writer. These were his brother William, Alexander Ireland, Robert Cox, and his wife. It was only at the publication of the twelfth edition that anonymity was dropped and it was finally revealed that it was Robert Chambers who was the author (Figure 3.12).

The scope of vestiges was broad, as it needed to be to fully develop his idea of a cosmic theory of transmutation. His intellectual scope was not confined to the biological world and handled all subjects as being part of an evolutionary system. On this plan, our solar system was formed from a cloud of gas which was created around our sun. Further, all life including man, and the Earth as well as the rocks that made it up, all evolved from the consequences of the developing solar system. Life was created, or arrived, by spontaneous generation. To back this up, he cited some experiments that insects could be generated by electricity, as a hypothesis this was quickly discredited. It was partly this use of extreme examples, which could not be justified or believed, that made *Vestiges* so controversial. Some of his ideas were of interest, paving the way in the popular mind for the work of Darwin. Chambers uses the fossil record to demonstrate the progression of simple to complex life, and ultimately man. Even though there was an implied idea of evolution, the activities and input of God were still assumed and acknowledged.

In the tenth edition of *Vestiges of the Natural History of Creation,* published in 1853, there is an illustration of a mammoth that would become one of the most widely used and republished illustration of a mammoth skeleton. The skeleton that the image is based on is still on show in St Petersburg, Russia. The illustrations are all very distinctly the same, even if on at least one occasion it is turned to face the other way (Hammerton 1906). What makes the picture so easily identifiable is that the skeleton came from permafrost and still retains skin and hair on the feet and skull. The original seems to have been produced in 1815 by Tilesius for the St Petersburg Academy of Sciences. After this, it reappears in various copies and redrawn lithographs. Some of the reproductions vary in quality, but appear in Cuvier (1825), and in a guide to the British Museum (Natural History) (Woodward 1890). Probably, the reason for the popularity of this image is that, like the skeleton itself, it is very striking. This is as much to do with its completeness as anything else, as in many of the associated captions, the remnants of skin is not mentioned. Although the anatomically detailed illustration that has been widely used is of interest (Figure 4.19), the issue with the tusks is well demonstrated by a picture of the same reconstructed skeleton that appeared in *The World Before The Deluge* by Louis Figuier (1863) Originally titled *La Terre avant le deluge* (Figure 6.2).

Chamber's book, *Vestiges*, was significant in stimulating discussion, while simultaneously pushing as fact, what to modern minds constitute some very peculiar ideas. His theories on development were based around a form of ontogeny recapitulating phylogeny. In his concept of the world as it was seen, species were real, but the relationships between them only vaguely considered. Changes from species to species were jumps, or saltations, not based on an evolving population, but on an individual. In this way, development and patterns of development were preordained; if an organism had a prolonged period of development, then it moved to the next step in what was also seen as a preordained hierarchy. This hierarchy was part of the fabric of design, as laid down by a divine creator. By having it as an

XXVI.—Skeleton of the Mammoth in the St. Petersburg Museum.

Figure 6.2 *The illustration of the St Petersburg Mammoth, from* The World Before The Deluge, *showing the outward facing tusks. This is the same animal as appears in Figure 4.19.*

embedded system within biology, it could be viewed in the same way as a law of physics. As such, once started, it could be left to run without any further divine intervention, leading to the ultimate goal. Chambers also had ideas of a higher power which could still instigate sudden change as though miraculous, although they would, in his system, constitute preordained shifts of emphasis and action.

One of the aspects of *Vestiges* which set it against many moral thinkers of the day and especially the religious caucuses of the church was the attitude of Chambers to mankind. The implication was that the culmination of creation, and all the manifestations of life in the mean time, was humanity. The problem was that this simultaneously required the belief that man was indeed, just another animal. As another species, a continuity of animal life was implied which did not single out man as anything other than another step in Chambers' unified hypothesis. He used a specific comparison to support his idea that while the inorganic world is entirely dependent on gravity, the organic world is entirely dependent on development. This is also a good example of the sometimes impenetrable reasoning that can be found in *Vestiges*. The book by Chambers was severely criticised on moral and factual grounds, the moral from clerical sources as a heresy, but more importantly from a factual point of view, by geologists. The geological sleight of hand that Chambers employed would have been believed by many readers. This was primarily in the form of an assumption of a continuous fossil record which could be used by Chambers to back up his ideas. Of course, then as now, the fossil record is incomplete and generally accepted as such. Obviously, the gaps were more so in the nineteenth century than the twenty-first, but there are still gaps, implying sudden leaps in biological organisation, which remain difficult to explain. Many of the changes in form and function as shown in the palaeontological record may well remain subject to considerable debate as evidence in favour of one hypothesis over another waxes and wanes, as in the case of the origin of flight. Flight not only requires structures which can sustain flight, but it also requires the unfossilisable, in the form of behaviour. This is different to the origin of feathers, where it can be expected that with time, a clear idea of where, if not when, feathers started their journey from reptile to bird. Flight, however, may be a by-product, or a primary result of feathers, but to be clear on that point, we may have to extend palaeontology from the material to the behavioural, which, of course, is fraught with assumptions.

The known gaps in the fossil record were of particular interest to Darwin, as he realised that they could be used to undermine his theory of evolution. Even so, the ideas as put forward by Darwin were still compatible with the fossil record, as it was known and including its gaps. In contrast, the ideas put forward by Chambers were brought into stark relief because he stated the fossil record to be complete, while it was clearly demonstrated not to be so. The slow unearthing of new fossils which gradually filled in some of the gaps, even though there was always some still remaining, aided Darwin, while emphasising the misconceptions Chambers used.

What Chambers did manage to create was a feeling that direct intervention by a deity was not necessary if speciation governed by law was accepted as an alternative hypothesis. This idea of a law, even one set up by a supreme being, sowed the seed of an idea that species were in some way interconnected, rather than individual creations. After all, inherent in the idea of species being created individually was the assumption that even if the body plan was similar, they were unrelated. If the assumption that every species was newly minted, it followed on, that any taxonomy was an artificial construction by man. The idea of a law set in train by a deity was not entirely without some interest, and it certainly gave some flexibility to the otherwise rigid ideas of divine creation. This was not, however, evolution and neither was it Lamarckism, which in itself had been seen as radical as it also introduced flexibility into the whole concept of species. It was into this maelstrom of ideas on the origin of life and the sanctity of species that Darwin published his seminal work *Origins* in 1859.

The long-term arguments over the work of Darwin fall broadly into three groups. The first is, of course, support, seeing it as a rational and materialist explanation of the observed world, from the fossil record through to contemporary species. The second strand of argument was against the view of Darwin on the basis that it was so materialist by its very nature that it refuted the existence of a deity. The third line of argument is often overlooked, since it is not based on anything but a political rhetoric. This is the idea that any scientific movement is a reflection of the social ethos in which it is formed. On this basis, Darwin's theory of evolution supposedly reflects the competitive nature of Victorian life, where commerce and manufacturing reigned as the great wealth maker and creator of freedom from poverty that was desired by all. Although this was an over simplification of the nineteenth century, it seems to have a parallel in the socio-political systems holding sway at the end of the twentieth century. However, as can readily be seen, it is less that Darwin's evolution came out of the society in which he lived, as society reflected the theory. By taking it as a simple explanation of everything, it imbued Darwin's ideas with an influence beyond biology. By forcing the social system to fit his ideas, it becomes possible to think of current models of business systems as being Darwinian. But be very careful with such simple analogies, this theory of such power is to do with a biological system, only. If it is pushed too far, we may as well use a train travelling along rails, not as a metaphor for electricity travelling along wires, but as a literal explanation, which, of course, it is not.

When socio-political disagreements with Darwin get out of hand, it can not only distort scientific progress, but worse, it can also adversely damage individuals. Such damage may not be to their career, but to the health and welfare of whole populations. Such a case occurred when Trofim Lysenko (1898–1976) gained political ascendancy in Stalinist Russia. Lysenko took the core idea of Lamarkism and extended it using political power into an entire system of biology. Lysenko started by rejecting Mendel's rules of inheritance. But he also rejected the idea of genes and later, the position of DNA in inheritance. After he had become head of the Institute of Genetics in the USSR, Lysenko enforced an anti-mendelism rule, dismissing scientists who adhered to the ideas of what we would now think of as classical genetics. While claiming authority in genetics, he seemed to have made up his ideas with no regard for evidence, such were his claims that they are generally regarded as pseudoscience. It was his misplaced notions of modified inheritance that was responsible for exacerbating an already bad series of famines under the agricultural collectivism. Even so, there has been a strange resurgence in interest in the ideas of Lysenko (Kolchinsky et al. 2017). The apparently new interest in Lysenko and Lamarck is in part associated with a misguided belief that epigenetics in some way validates the older and incorrect ideas. This is nearly always because the protagonist does not understand epigenetics and has little understanding of Lysenko. Notwithstanding the waxing and waning of Lamarckism, it frequently creeps into evolutionary ideas by mistake. For example, during the nineteenth century, it was a common misrepresentation to talk of evolution in terms of the neck of the giraffe extending through generations of stretching for high leaves.

While the ideas of Darwin gained considerable traction in creating a viable explanation for evolution and extinction, there were often over enthusiastic interpretations put forward as fact. For a period during the nineteenth century, it became possible to see various attempts to integrate older ideas with the new concept of natural selection. Part of the impetus for this composite evolutionary cookery was the perennial problem of an incomplete fossil record. To create a credible theory, it was necessary to fill the gaps with something else, so in place of unknown fossils, embryology took its place. The master of this idea was Ernst Haeckel (1834–1919) who was a champion and populariser of the concept of ontogeny recapitulating phylogeny. Although interesting as an idea, it did tend to become an accepted fact, a biological law, even. By uncritical use of this, it did slow down sensible use of palaeontology to demonstrate evolution and help explain the history of the world. During the period that ontogeny and phylogeny were inextricably mixed, the emphasis on fossils as guides to ancient relationships became lost, or at the very least, reduced to a subservient position in studies of the evolution of life on Earth. This is shown in the two volumes of *The History of Creation* (Haeckel 1876), which bases the whole argument of phylogeny on living species. Consequently while the title is misleading, these volumes remain extremely good and clear works on embryology and developmental processes in the animal kingdom.

Almost as an essential component of the development, reflecting phylogenetic relationships argument is that, if true, it can only show a linear progression, which it is reasonable to assume is not necessarily the case. Similarly, if a developing embryo reflected past complete life forms, then surely, by extension, it must be the case that extant species that still exhibited these early characteristics were in some way 'lower' species, less developed or less advanced. This negated all of the previous ideas that had been pivotal in the early understanding of ecological systems where it was becoming clear that all species are adapted to their environment to survive in their ecological niche to guarantee survival. Haeckel took the next step by assuming that there was some form of progression which by the logic of the day would require that Man was the ultimate outcome. It was this idea that generated the famous illustration of a tree of life (Haeckel 1897) with man at the top and all other species lower down.

Using the linear model of evolution would allow for prior knowledge of the forms of fossils that would be expected to be found. The form of fossil species, on this model, remains unchanged through vast periods of time, becoming simultaneously both the ancestors of modern species and themselves living fossils. Because the fossil record was recognised as incomplete, it became a normal assumption that not finding fossil remains of these earlier species was not an indictment of the linear model, but a reinforcement of the fragmentary fossil record. As a logical argument, this lacks any scientific rigour, but was constructed before a clear idea of statistical analysis of data was available. The linear model, although accepted as an easy idea amongst biologists, was at odds with the branching evolutionary system of Darwin. Using a branching evolution model, it would be expected that modern species, adapted as they are, to modern ecological conditions would not be the same as their predecessors. It would also imply that species did not arise from each other but from a common ancestral form, which would not be expected to resemble currently extant species. Haeckel was also the originator of the idea of a temporal phylogenetic tree (Figure 6.3), rather than the tree of relationships which had been in current usage until that time. Although the difference between the two may seem minor, by introducing a temporal element into the diagrammatic tree, it became automatically implied that the relationships were linear. This added to the 'non-branching' idea of evolution, carrying as it did the authority of an established biologist. What seems to have gained considerable ground during the nineteenth century was as much a philosophical idea of change not requiring a theistic component, as a clear idea of evolution as a statement of what happened to biological clades over time.

During the nineteenth century, some of the gaps in the fossil record were recognised as so vast they strained evolutionary theory. The most notable of these involved the origin of birds. It had already been pointed out by Huxley that it was sometimes difficult, even impossible, to distinguish the feet and footprints of small dinosaurs from modern birds. While this could be seen as an indicator of lineage, there is a large conceptual gap between making such a suggestion and having evidence for it. Fossil evidence for the link came at a time when gaps in the fossil record were simultaneously pounced on by those who did not want to accept evolution, and those that thought gaps indicated simple difficulties of finding

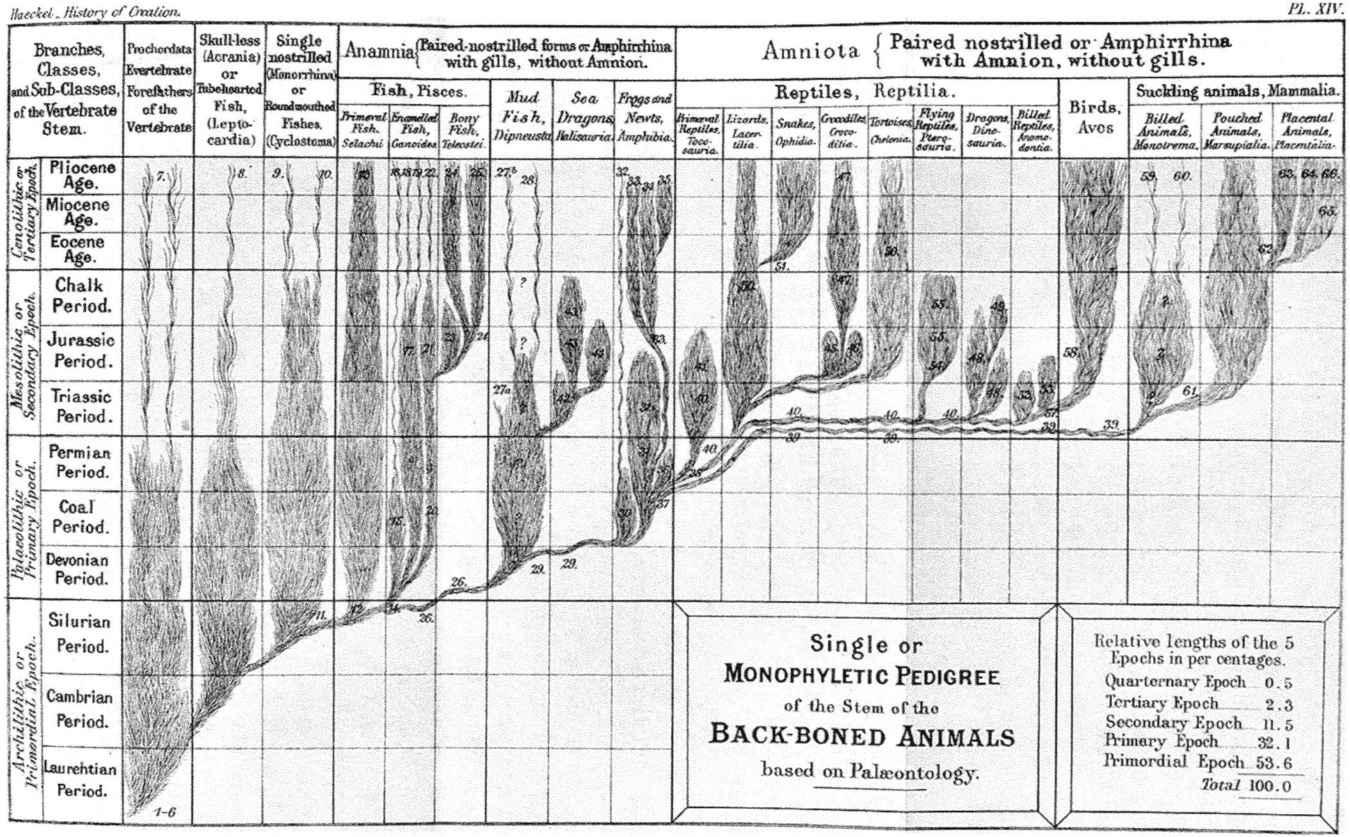

Figure 6.3 The temporal phylogenetic tree introduced by Haeckel. He used it in many different forms in different publications, this particular illustration is from History of Creation (1876).

fossils. So when, in early 1861, a feather was discovered in ancient deposits it was seen as pivotal to avian palaeontology. The first of several fossils of feathered forms, the feather was discovered in a quarry located in Bavaria. The feather was found at Solnhofen imprinted into the fine grained limestone of the Upper Jurassic, Middle Kimmeridgian, which has such a fine structure that it was routinely used for lithography and gained the generic term Lithographic Limestone. It was a deposit that had a long history of fossil discoveries, but the discovery of a feather was something quite different to those that had been seen previously (Swinton 1965). That first feather was described in 1861 by Meyer as 68 mm long with a vane 11 mm wide. Having been split, on the slab, there was a fossil and an impression. The main piece, with the fossil, went to Munich and the impression went to Berlin. It has been demonstrated since then, that this is not an avian feather of modern construction.

It was also in 1861 that the first skeleton was found at Langenaltheimer near Papenheim in Bavaria, which was intimately associated with feathers, this turned out to be *Archaeopteryx*. All of the specimens of *Archaeopteryx* so far discovered have been from Bavaria and found in the high-resolution lithographic limestone quarries. It was said that with this discovery, even the workmen, long used to coming across fossils, were astonished. It was apparently a reptile, but with clear indications of feathers on some parts. This first discovery caused a sensation across Europe and was described and named by Meyer as *Archaeopteryx lithographica*. It did not take very long for commercial interest to be raised, and by February 1862, the Natural History Museum in London was in contact with the owner, Dr. Häberlein. The fossil was bought by the museum for £700 and arrived in good order in October 1862 (Figure 6.4). It is said that the money became the dowry for the daughter of Dr. Häberlein. Since that first discovery, there have only been a further 12 found and no doubt there will be more as time passes and the quarries are either continually worked for their stone, or become derelict sites and treasure chests for palaeontologists. Even with so few specimens, it has become accepted that these skeletons represent two species, both in the genus *Archaeopteryx*. These are *lithographica* and *siemensii*. *Archaeopteryx siemensii* was described from one of the most complete specimens discovered and was described in 1884 as a new species. It was for some time regarded as a synonym of *lithographica*, although now regarded as a different species in its own right. It took its name from Ernst Werner von Siemens who funded the original purchase of 20 000 Gold Marks for the Berlin Natural History Museum. To put this in perspective, at the time £1 was worth about 20.43 Gold Marks and the Governor of the Bank of England was paid £400/annum. It was this same Siemens who was the founder of the eponymous company that still trades today. Because there are so few Archaeopteryx specimens, most of them are described by names which reflect either where they are kept, or where they were originally found.

(a)   (b)

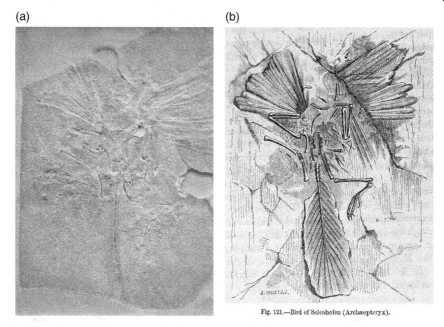

Fig. 121.—Bird of Solenhofen (Archæopteryx).

Figure 6.4 (a) *The British Museum* Archaeopteryx *from Bavaria.* (b) *An illustration of the fossil, by Figuiera in* The World Before the Deluge *(1863).*

While there has been controversy surrounding these fossils, two things have been consistently agreed upon. The first is that it is a bird and the second is that it is very closely allied to the reptiles. This co-mingling of characters in a single fossil was what had been hoped for by evolutionists. It was certainly in direct contradiction of the ideas of papal literalists for whom the characteristics which defined phylogenetic groups should not cross boundaries. The finding of *Archaeopteryx* demonstrated that such taxonomic ideas could not be projected backwards through geological time and that here was a found, rather than missing, link between two classes. This was particularly reinforcing for Haeckel who had described an evolutionary link between birds and reptiles based solely upon the embryological development of birds being apparently so close to that of reptiles. The discovery and acceptance of *Archaeopteryx* as a fundamental link between reptiles and birds made all enquiring minds think twice before any link between two apparently disparate contemporary groups was disregarded out of hand. If reptiles and birds could be linked in this way, then it became possible to think of a range of relations between fossil forms that could go a long way to elucidate evolution.

A similar and in some ways more complete story surrounds the fossil records of modern horses. This story does not have the same significance in the popular imagination as *Archaeopteryx*, but in the investigation of the processes and outcomes of evolution, it is at least as important. The gradual evolution of horses, with its human intervention in historical time as well, lacks the extraordinary concept of the jump from land-based animal to birds soaring in the air. What it does have is a well-regarded and understood range of fossils showing the way that the single equine toe developed.

This was also a story which, as it unfolded, was to influence all of palaeontology. It helped direct the study of the subject towards evolution and elucidating the relationships. Palaeontology was no longer just about recording finds; it was about studying relationships not only between fossil forms but also between fossil forms and extant species. There were a series of ungulate fossils which were thought to be related, but at the start of the nineteenth century, precisely how, was unclear. In 1871, Vladimir Kovalevsky (1842–1883) looked again at the fossil that Cuvier described as *Anchitherium*. The conclusion he came to was that this was a direct link between *Hipparion* and the much older *Palaeotherium*. At the time, *Palaeotherium* was considered to be an antecedent of the modern horse lineage. Now it is recognised to be a sister lineage with a much more ancient common ancestor. The search for a palaeontological family tree for horses inspired O.C. Marsh (1831–1899) to investigate the fossil record of his home country, the USA. This would prove to be highly productive and decisive in helping to understand the lineage of *Equus*.

Using fossil material from the USA, Marsh began to form a more complete picture of changes to the equine clade over geological time. His work produced an almost unbroken lineage from the Eocene *Orohippus* through to modern *Equus*. He published his results in 1874 (Marsh 1874). The illustration which accompanied the later publication (Marsh 1879) has been reproduced many times showing the reduction in toes from the earliest form through to the single toed modern horse (Figure 6.5). There are also diagrammatic representations of the teeth and leg bones, but it is undoubtedly the fore and hind feet images which give the clearest idea of the transition from the earliest form to the most recent. Marsh also predicted the possible form of an ancestral type, the 'Dawn Horse' which he called *Eohippus*. Although this was a speculation, it was proven correct quite soon afterwards with the discovery of a fossil which fitted the requirements, although it was known as *Eohippus* before it was known from remains, the fossil species is now called *Hyracotherium*.

Marsh represented the evolution of the perissodactyls primarily based upon fossils from the USA, with the fossil record, incomplete at best, of Europe being shown as side shoots from the main line. Huxley even assumed that the fossils that had been found and described from Europe were mainly those of occasional migrants. It was one of the best argued and illustrated accounts of evolution as explained by the fossil record which had been made. The illustration which is best remembered from the work of Marsh, that of the 1879 paper, did have the

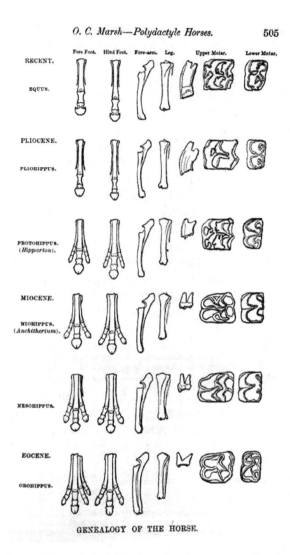

O. C. Marsh—Polydactyle Horses. 505

GENEALOGY OF THE HORSE.

Figure 6.5 *Evolution of the foot of modern horses as described by* Marsh (1879).

consequence, by virtue of its layout, of implying a linear progression from one species to the next in a smooth progression. The real truth would be more like a series of branches with common ancestors at the axial points of divergence not necessarily being represented. Marsh also inadvertently moved American palae-ontology from being a process of collecting and describing, towards an under-standing of fossils and reasoned speculation based upon good scientific evidence of their significance.

Prior to the nineteenth century, the question of Man's origin had been largely avoided, religious scriptures of all sorts had a monopoly on that. During the nineteenth century, this changed and the origins of Man became a legitimate subject of study. This was one of the aspects of Chambers' work *Vestiges* that caused such consternation and vituperative criticism. Although it was frequently misinformed and incorrect in its statements, *Vestiges* was perceived by the clergy and laity as undermining the status of humanity. It managed this by invoking a naturalistic approach to creation which just did not work with a theistic intervention. That the work by Chambers raised considerable ire can be seen in the publication in 1849 of an attempt at rebuttal by Hugh Miller (1802–1856) (Miller 1849). This was called *Footprints of the Creator,* with a subtitle of *The Asterolepis of Stromness.* The *Asterolepis* in the subtitle refers to the extinct genus of heavily armoured placoderm fish and reflects the detailed knowledge that Miller had of fossil fish from the Old Red Sandstone of the Devonian. He used this knowledge in his book to undermine one of the central arguments of *Vestiges* by demonstrating the claim of Chambers that these very early vertebrates were less complex and less adapted than modern fish was incorrect. Miller was also using his work as a method of claiming that creation was the direct result of action by a benign creator.

The use of this argument and others by Miller was specifically to demonstrate the existence of a benign creator. This in turn was explicitly to negate the argument that if Man had appeared by chance, as Chambers suggested, then philosophically he could not be held responsible for his actions. It was this argument that was perceived by the polite and educated of the nineteenth century to be likely to undermine the social fabric that held society together. In some quarters, this was regarded as even more of a threat than open revolution. This was a heated debate as much for the innate curiosity of people regarding their origins as for the social implications. Nonetheless, the arguments would be examined and re-examined as to what the palaeontological lineage of man was on one side, or what sort of a creator would do this to the planet on the other side.

There was one major aspect of the origins of humanity which both sides, the Divine and the Natural, used as propaganda. This was the lack of any fossils which directly linked Man to animals. The imperfection of the fossil record had been acknowledged by Darwin with respect to ancient species of animals, but he had shied away from passing comment on this aspect of human evolution. Darwin had steered a path throughout *Origins*, which did not pass comment on the human condition. It was a feat of considerable self control by Darwin to limit his opinion to a single sentence 'Light will be thrown on the origin of man and his history'. This contained both the acknowledgement that fossil remains were as yet unknown and also that such a thing as the origins of man could not remain hidden indefinitely. It was almost inevitable that with so very few fossil remains of Man or human progenitors, every opinion as to the origin of man would find an advocate. These would be based upon the personal ideas and hypotheses of the individual, whose conviction would quite often outstrip their knowledge. From a purely objective point of view, with no available evidence, all suggestions are equally likely. For some, the determination of the truth was based around fantasy, the product of a blind and slavish belief in the Bible. For others, it was a strange interpretation of the fossil record of 'lower' forms, combined with a leap of faith when sound information ceased, and the easiest way of explaining humanity was to claim at that point a special status for Man.

There were many who considered man to be the ultimate result of evolution, or the ultimate aim of evolution, or the final product of Creation. These were all widely held ideas and had the common point that man was on top of the tree of life. Such was this perception that Haeckel, in his work *Evolution of Man* (1897), has an illustration of the Pedigree of Man, which is illustrated as a tree (Figure 6.6). It is this illustration which has been reproduced, copied and interpreted in print in innumerable publications since the end of the nineteenth century (Figure 6.7). This illustration clearly puts man at the very top of the evolutionary tree, with all the other various animal groups as offshoots from the main trunk. The overall impression is one of animal life all leading to the production of man; it gives no hint or credence to ecological interactions or independent adaptation to the environment. The image, although from Haeckel who was keen on his version of evolution as a concept, gives credibility to a teleological version of evolution with a goal in mind. This image of a tree of life was going to have far reaching consequences because although simple to understand, it is recognised as wholly wrong. It is perhaps judging Haeckel unfairly to put too much importance on this one image, produced as it was in the nineteenth century. As he says quite explicitly at the start of *Evolution of Man*, the increasing popular interest in the other sciences induces him to try to do the same for the origin of man. Although he hopes to succeed, as he admits, 'it is in many respects especially beset with difficulties'.

Haeckel reproduced another illustration which has become rather misunderstood and misinterpreted. This was used originally as the frontispiece to Thomas Huxley's work *Evidence as to Man's Place in Nature*, published in 1863. In this book, he gives his evidence for the evolution of man and apes from a common ancestor. This included the implication that evolution applied to man as much as any other organism, which was an uncomfortable idea for many of the increasingly well read middle classes, whether or not they were particularly religious or not. The Frontispiece, which showed the supposed progressive changes from gibbon to man, was originally a series of engravings by Waterhouse Hawkins which were rescaled for affect, with the gibbon being twice scale size. Haeckel seems to have re-used the image in *Evolution of Man*, but without acknowledgement to the original (Figure 6.8). The way which has become misunderstood is in the use of it in pastiche form. This reproduction of it in different forms came to its zenith in

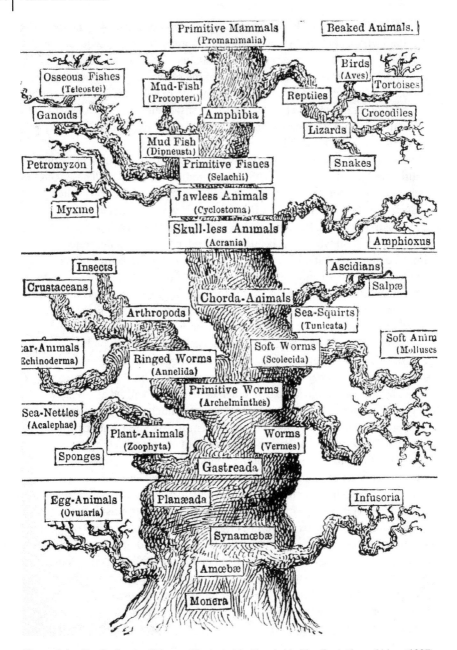

Figure 6.6 *The Pedigree of Man, as illustrated by Haeckel in* The Evolution of Man *(1897).*
Source: Haeckel.

Figure 6.7 *The tree of life as it appears in the end papers of* The Miracle of Life *(1939). by H. Wheeler. This popular science book made modern man as the pinnacle of evolution in this pictorial representation* (Wheeler 1939). *Source: Wilson J Wall.*

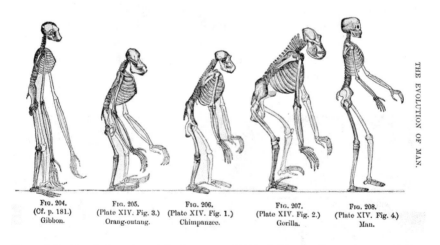

Figure 6.8 *Frontispiece that originally appeared in* Man's Place in Nature (Huxley 1863) *and was copied for* Evolution of Man (Haeckel 1897), *where this illustration comes from.*

Early Man (1965) published by Time Life Books the illustration was drawn by Rudolph Zallinger, which was a series of 15 images from rudimentary humanoid to modern man. Each human image becomes larger, taller, straighter, as the series progresses on its fold-out illustration. Since the appearance of this illustration, many pastiche versions of it have been produced, mostly for the garment industry where it appears on T-shirts (Figure 6.9).

Before the advent of palaeontological evidence, searches were made for any anatomical characteristic which specifically separates man from animals. This was paralleled by the continuous desire to make man a pinnacle of life, not necessarily evolution, to make humanity unique and responsible directly to the creator. The most significant and complete method of separating humans from apes started in 1779 with Johann Blumenbach. Although it was his original idea, it was quickly picked up and endorsed by Cuvier. The notion that Blumenbach put forward was that apes were four handed, that is hands on all limbs, which he called quadrumana and humans had two hands and 2 ft, hence they should be bimana. This was in many ways a distinction without a difference, but it did allow for an implied superiority for man over the apes. Huxley was not convinced by this apparently trivial difference in function for terminal appendages and argued that it was not significant enough to make a distinction, so both humans and apes should be regarded as Primates.

By having a single group of Primates, within which mankind sat, pulled the group together, but still relied on anatomy of extant species to supply the evolutionary link. In many other groups, this would have been sufficient. For taxonomists and palaeontologists, it was becoming increasingly clear, as well as in the popular imagination, that there was a fossil record which demonstrated origins for any group of organisms. It was also the functional similarities of the group which were reflected in surviving species. For the primates, this was just not seen as an adequate argument for a relationship. Before it was ever going to be accepted

Figure 6.9 *Two pastiche images of* Early Man (Rudolph Zallinger 1965). *Both of these appear on T shirts.*

that Man belonged within this group, hard evidence was going to be required. Although such evidence was a long time forthcoming, the lack of fossil evidence did not stop the arguments and debates regarding the origins of Man from being made.

Haeckel in *The History of Creation* (1876) used the lack of direct fossil evidence as a good reason to use other fossils as guides to the age of the human race. He tries to pin down the transmutation from 'man-like apes into the most ape-like men' by assuring us that man is derived from placental mammals, which only occur from the Tertiary onwards, so man must have developed, at the very earliest in the Tertiary period. Haeckel suggests that this step came at the end of the Tertiary, in the Pliocene, possibly even the Miocene. His primaeval man is referred to as alali, or speechless. This makes a considerable assumption, and although not affecting the basic reasoning of the origin of man, it has little by way of sound logic to back up the claim that humans, in their modern form, are the only ape, either antecedent or contemporary that has speech.

The continuing lack of a coherent fossil record for man throughout much of the nineteenth century was referred to by Lyell. He pointed out that it is a slow and lucky process regarding fossils of any sort to both form them and find them. What is more, since Lyell was a proponent of an African origin for modern man, the areas most likely to reveal fossils had not yet been properly searched. As a hypothesis, this site of origin was in contrast to Haeckel who was keen on the idea that man originated in south east Asia. In support of an Asian origin was an ape-like hominid, *Pithecanthropus erectus,* more commonly called Java Man. By 1909, about 11 different specimens had been found of Java Man, which pushed the probabilities heavily towards an Asian origin.

When Darwin published *The Descent of Man* in 1871 (Darwin 1871), he was already of the opinion that the idea of quadrumana and bimana was being abandoned in favour of a single group of primates. Darwin agreed with this change in light of the various anatomical difficulties associated with the classification. Although this was not explicitly stated, the definition would have been associated with the idea of an opposable thumb. The available observations of such dexterity would have been limited and a completely opposable thumb, able to contact all other digits, impossible to be certain of. This makes the difference between a foot and a hand contentious and any gradations of usage so subtle that it would make a primary taxonomic dichotomy impossible to support. Darwin goes on to suggest that there is a 'comparative insignificance for classification of the great development of the brain in man'. This is a well considered line of reasoning, since it is analogous to neck length in antelopes, of little phylogenetic or taxonomic value. Indeed, Darwin goes on to say that other differences are similarly merely adaptive in nature. He uses the parallel of seals, highly adaptive, but still regarded as a family soundly within the order Carnivora. The comment that Darwin makes regarding the mental capacity of humans as an indicator in some way of the superiority of man is really quite prescient. While most of the nineteenth century scientists would implicitly assume that man was in some way the pinnacle of evolution, the lack of acceptance of intelligence as a key factor in distinguishing humans from animals, as though in some way separate, is tacitly implied in the statement of Darwin. While the argument for regarding man as another animal is not addressed directly, it does remain in the background, probably as much to allow for evolution to have worked on man as anything else.

Darwin for many reasons preferred the idea of an African origin, rather than an Asian one, for humanity. This gave the zoological origin for man of as a common Catarrhine (the spelling of this varies between one and two 'r's) ancestor in the far distant past. This would be a common ancestor, but would not be a result of linear descent. What was hampering these speculations and investigations was the lack of fossil remains. Darwin was, nonetheless, adamant that the idea held by some of his contemporary naturalists that there were three Kingdoms, Human, Animal and Vegetable was wholly incorrect. It was, by his reasoning, inconceivable that the similarities between human anatomy and animal anatomy were coincidental. He was also of the opinion that the extreme mental capacity of man was exactly that, an extreme version of that found in animals. In one of his most lucid explanations reduced to a single sentence Darwin opined:

> If man had not been his own classifier, he would never have thought of founding a separate order for his own reception. *(Descent of Man 1871)*

This was echoed by Huxley, also quoted by Darwin as there 'is no justification for placing man in a distinct order.' This was all based upon the anatomy and physiology of the Catarrhine and Platyrrhine primates, even though there was some fossil primate material becoming available during the nineteenth century. The idea that man originated in Africa was at the time a speculation, although as was pointed out, the discovery of a fossil primate from the Miocene, *Dryopithecus*, in 1856 by Edouard Lartet in Europe did nothing to clarify the situation. Such vast lengths of time had passed before and after the Miocene, that there had been ample time for migrations to have occurred on a huge scale. The reflections on human migration at this time was an interesting one, since until the detailed investigations of the origins and distribution of horses became a question for palaeontology, the question of social activities and interactions had not been posed for fossil species of any sort. This is notwithstanding the pictorial representations of the world in previous times, these were vignettes, rather than serious suggestions of interactions and ecosystems (Figure 6.10).

During the nineteenth century, few human-like fossil remains had been found and those that had been were as much a source of confusion as a source of clarification

XVIII.—Ideal landscape of the Middle Oolitic Period.

Figure 6.10 *A vignette representing animals of a bygone era from* Figuier (1863), La Terre avant le deluge. *This was not intended to represent a true image of life at the time, as demonstrated by the original explanation for this illustration. On the beach is* Hylaeosaurus, *emerging from the water is* Teleosaurus cadomensis, *in its mouth,* Geotheutis, *a species of squid. To show the unusual underside of* Teleosaurus cadomensis, *one has been represented as dead and floating in the shallow water.*

of man's position in nature. While *Dryopithecus* was recognised as being related to the Hylobatids, (gibbons), other skulls were more controversial in their taxonomy. One of the first of these was the Engis skull which was found in 1829 by Philip-Charles Schmerling in Belgium. Schmerling is often described as a founding figure in palaeontology and was one of the first scientists to recognise the existence of prehistoric man. The Engis skull was not in any way complete, so trying to form a clear idea of what the living animal would have looked like, was virtually impossible. When the Engis skull was first discovered and described, it was regarded as fully human, although we know it now, to be Neanderthal. As the Engis skull and slightly later finds were assumed to be modern and not pre-human, they were regarded as of little interest in determining the origin of man. Some of the remains indicated a heavy brow, consistent with an ape-like origin, but some considered they were human but of pathological origin. To Huxley, the paradox was that it was both ape-like, but with a human cranial capacity. Huxley had always assumed that any link between apes and man would have a skull capacity intermediate between the apes and humans.

While the zoological origins of humanity were being investigated, so too were the geographical origins. Although there was much to suggest that mankind had its origins in Africa, there was still considerable feeling that Asia may have been the cradle of humanity. This was championed by Haeckel, though he was not alone in this. The emphasis on Asia was given a considerable impetus with the discovery of *P. erectus*. The original discovery was made in Java by Dr Eugene Dubois who was working for the Dutch Army Medical Service. Although not the first explorer to make such a declaration, he went to East Java with the expressed intention of discovering remains of early man. He searched between 1891 and 1894 in Java, with the breakthrough coming in 1892, the year when Dubois started excavating a bone bed near the village of Trinil, positioned on the Solo River half way across Java between the Java Sea and Indian Ocean. This was not a random choice on his part as bones of extinct fauna had already been found along the banks of the Solo River, the burying deposits being layers of volcanic ash. His original find was what appeared to be a human tooth, later followed by the top of a skull and femur as well as other teeth. This collection of specimens quickly became known as Java Man and changed the consensus ideas on sites of origin for humanity (Figure 6.11). Subsequent German expeditions to East Java made several new discoveries of animal bones, but nothing more of *P. erectus*. The skull cap

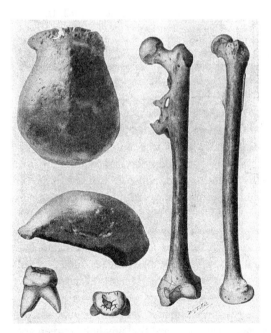

Figure 6.11 *The major components of the original finds of* Pithecanthropus, *discovered by Eugene Dubois. This composite illustration is of the top of the skull, in profile and from above, femur front and profile and third upper right molar, from above and in profile.*
Source: https://commons.wikimedia.org/wiki/File:Pithecanthropus-erectus.jpg#/media/File:Pithecanthropus-erectus.jpg, Public Domain.

was the most important find, but it remained controversial as it was far more primitive than anything that was distinctly human in form, such as Neanderthals. The femur by virtue of its shape indicated an upright, human, posture. Dubois was inclined to consider that this may be an intermediate between ape and man, an ape-like skull and a human posture. Although originally considered Pliocene, the geology indicated that it was Pleistocene. In the 1920s, these original excavated relics were investigated more fully, along with more material of the same sort. These also came from the valley of the Solo River, although not from the same sites. With the collection of remains, mostly of skull, jaws and teeth, it became possible to assess *Pithecanthropus* as a man, rather than an ape.

This important discovery was in part a result of Haeckel influencing Dubois and his choice of site to investigate, but it also reflected a lack of palaeontological field work on the continent of Africa. This very much comes down to, if you do not look for something, it will not be found. As a consequence of this, those places where evidence was looked for and that produced data, would automatically become the assumed centre of human origin. The search for fossilised human remains was, in this way, still dominated for many years by the availability and access to suitable research sites. In many geographical areas, discoveries were frequently made, moving the fundamental question of where we come from forward, in small incremental steps. One such successful expedition from Sweden was made to China in the 1920s, to investigate the limestone caves of Zhoukoudian. At the time of the expedition, the normal Anglicisation of Chinese names was the Wade-Giles system, and so this town was called Choukoutien. This is a site a few kilometres south west of Beijing where the caves contained considerable deposits of animal remains. In amongst this collection, Dr. W. Pei found two teeth. These were some sort of primate, but whether ape or man was initially unknown. At the medical school in Beijing was an anatomist, Davidson Black, who concluded after studying a third tooth found at the site, that it was from a primitive human. Based on this evidence alone, Black published a description of a new genus of human, given the name *Sinanthropus pekinensis*. Two years later, in 1929, a cranium top was discovered from a primitive human that was thought to have come from the same species as the teeth. Over the proceeding years, several more cranial pieces were recovered, along with various limb bones. Overall there is a close similarity between *Pithecanthropus* and *Sinanthropus,* but Peking Man, *Sinanthropus*, has a larger cranial capacity than Java man and was already a tool user with napped stone implements being found in close proximity and the same strata to the remains. In the same area and of the same geological age, there were several hearths, implying that *Sinanthropus* was a fire user. It is unfortunate that the soriginal fossils went missing from a train that was to carry them to the coast, where they were to be shipped overseas. *En route* to the ship, the train was captured by Japanese troops and from that point the collection of exhibits were lost.

As a result of these finds, it was long considered that Asia was the place that mankind had originated. This situation changed in 1925 when Dart published his description of an African species from Taung, about 130 km north of Kimberley in South Africa (1925). The discovery of these fossils came from limestone tufa being quarried for cement. In part of the cliffs, there were caves that had been back-filled over time and in the filling there were many fossils of Pleistocene mammals, mainly baboons, but including a partial anthropoid skull that was different from any other fossil form. Named as *A. africanus*, it was not without controversy as some palaeontologists questioned whether it was any sort of hominid. In the same area, there were other remains that were clearly not from *A. africanus* and were at first named as *Paranthropus robustus*. It was later realised that although different, this second species was still Australopithecine, and so was renamed *Australopithecus robustus*. Although the work by Dart raised a genuine question regarding the geographical origin of humanity, there was an increasing number of early hominid fossils originating in parts of Asia which kept the idea of an Asian origin alive and well.

The movement of an evolutionary origin for mankind towards Africa was enhanced with the demonstration from pelvic remains that *Australopithecus* was most likely to have an upright, bipedal posture. This was something of a surprise because of the relatively small brain size of about 450 cm^3. The two suggested species of *Australopithecus,* that is *africanus* and *robustus*, seemed to live in close proximity and would have had a common ancestor. However, *A. robustus* was seen as an offshoot, while *A. africanus* was seen as a potentially direct progenitor of *Homo*. Development of ideas regarding human origins was considerably altered when L.S.B. Leaky published his finding on a new species which he named *Zinjanthropus boisei* in 1959 (Leaky 1959). This had been found in Olduvai Gorge in Tanzania and was very close, but not identical to *A. robustus*, consequently its taxonomic position was adjusted and it was renamed as *Australopithecine boisei*.

The evidence for an African origin was increased when remains very close in structure to *P. erectus* were found in Africa, suggesting a common lineage with the Astralopithecines. A review of the genus by Mayr (1963) suggested that the difference between this and modern man was more akin to the differences between species, rather than genera, so he renamed the fossil *Homo erectus*. Further discoveries within the genus *Homo* resulted in the earliest species of the genus, with a cranial capacity of 600 cm^3, being found. *Homo habilis* lived about 3 million years ago was acquainted with fire and was already a tool user. There is also distinct evidence that *A. boisei* and *H. erectus* coexisted around lake Turkana as the remains are at the same stratigraphic level. Whether there was ever any interaction between these groups remains unknown.

A significant development in the ideas around human evolution came with the discovery of many fragments of bone from a single individual, which was formerly named *Australopithecus afarensis* and informally became known as Lucy.

As the cranial capacity was no more than 500 cc, this helped to clarify the idea that bipedalism predated an enlarged brain. Lucy was dated to about 3.2 m years ago. Further evidence of this gap between bipedalism and brain size came with the discovery in 2002 of *Sahelanthropus tchadensis* in the African state of Chad. This primate, in the family Hominidae, had a brain case of about 350 cm^3 and a foramen magnum which would be consistent with bipedalism, although no direct evidence regarding the stance has been found. At about 6 million years old, this would make it one of the earliest bipedal apes, should it turn out to be such. The foramen magnum is a good indicator of how the head of a mammal is held, and in bipedal primates, it is underneath the skull, giving a direct line from the spine to the brain, with the eyes pointing forwards.

Although there was a period in the nineteenth century when postulating about the origin of humanity was based around physiology and anatomy of extant species, there was little by way of fossil evidence to contradict any suggestions made. With time, fossil evidence was found from many different sources, with implications for the geographical origin of mankind. Something else was discovered from the fossil evidence; that bipedalism and a large brain were developed separately. This has considerable implications for *Homo,* the development of a large brain independently of any other feature suggests that this is not an inevitable result of evolution. For the vast period of geological time during which the Earth has been in existence, the ecology has not been dominated by intelligence, but by a network of interacting species creating a balanced ecology. It would appear that an intelligent animal is neither inevitable nor essential in a living system, which should help to revise evolutionary trees away from having *Homo* as the apex species and towards a more integrated expression of life.

References

Agassiz, L. (1833–1843). Recherches sur les Poissons Fossiles, Five volumes.

Bateson, W. (1901). Experiments in plant hybridization. *Journal of the Royal Horticultural Society* 26: 1–32.

Boule, M. and Vallois, H. (1957). *Fossil Men, a Textbook of Human Palaeontology*. London: Thames and Hudson.

Bowler, P.J. (1983). *Evolution: The History of an Idea.* Berkley: University of California Press.

Brongnart, A. (1822). *Sur le Classification et la Distribution des Vegetaux Fossiles.* Paris: De A. Belin.

Buffon, G.L. (1778). *The Epochs of Nature* (trans. J. Zalasiewicz, A. Milon, and M. Zalasiewicz). Chicago, USA: University of Chicago Press.

Butler, S. (1872). *Erewhon.* London: Trubner and Ballantyne.

Butler, S. (1979). *Evolution Old and New: Or the Theories of Buffon and Erasmus Darwin and Lamarck as Compared with that of Mr Charles Darwin*. London: Hardwicke and Bogue.

Chambers, R. (1844). *Vestiges of the Natural History of Creation*. London: Churchill.

Cuvier, G. (1825). *Researches sur les ossemens fossiles de quadrupedes*, 3e 1825. Paris.

Dart, R. (1925). *Australopithecus africanus*: the man ape of South Africa. *Nature* 115: 195–199.

Darwin, C. (1859). *On the Origins of Species by Means of Natural Selection: Or the Preservation of Favoured Races in the Struggle for Life*. London: John Murray. The various editions published during Darwin's lifetime have additions and revisions. This means the many modern reprints which are available will vary depending upon which edition they replicate.

Darwin, C. (1871). *The Descent of Man*. London: John Murray.

Darwin, C. (1880). Sir Wyville Thomson and natural selection. *Nature* 11 (23): 32.

Eiseley, L. (1958). *Darwin's Century: Evolution and the Men Who Discovered It*. New York, USA: Doubleday.

Figuier, L. (1863). *La Terre avant le deluge*, Paris. Translated Henry Bristow (1867). *The World Before The Deluge*. Chapman and Hall, London.

Haeckel, E. (1876). *The History of Creation*, vol. 2. London: Henry King & Co.

Haeckel, E. (1897). *The Evolution of Man (2 Volumes) D*. New York, USA: Appleton and Company.

Hammerton, J.A. (1906). *Harmsworth Encyclopaedia*. Eight Volumes, vol. 5, 3970. London: The Amalgamated Press.

Huxley, T.H. (1863). *Evidence as to Man's Place in Nature*. London: William and Norgate.

Kolchinsky, E., Kutschera, U., Hossfeld, U., and Levit, G. (2017). Russia's New Lysenkoism. *Current Biology* 27: 1042–1047.

Lamarck, J.B. (1809). *Philosophie Zoologique*, vol. 2. Paris: Dentu.

Leaky, L.S.B. (1959). A new fossil skull from Olduvai. *Nature* 184: 491–493.

Lyell, C. (1832). *Principles of Geology*, vol. II. London: John Murray.

Marsh, O.C. (1874). Notice of new equine mammals from the tertiary formation. *American Journal of Science* 7: 247–258.

Marsh, O.C. (1879). Polydactyl horses, recent and extinct. *American Journal of Science* 17: 499–505.

Mayr, E. (1963). The Taxonomic evolution of fossil Hominids. In: *Classification and Human Evolution*, Publications in Anthropology, vol. 37 (ed. S.L. Washburn). New York: Aldine Publishing.

Mendel, G. (1866). Versuche über Pflanzen-Hybriden. *Verh Naturforsch Ver Brünn* 4: 3–47.

Miller, H. (1849). *Footprints of the Creator*. Edinburgh: William Nimmo.

Roger, J. (1989). *Buffon: Un Philosophie au Jardin Du Roi*. Paris: Fayard.

Swinton, W.E. (1965). *Fossil Birds*. London: British Museum (Natural History).

Thomson, S.W. (1877). *The Voyage of the Challenger*. London: Macmillan & Co.

Wallace, A.R. (1879). Book review of *Evolution Old and New*. *Nature* 20: 141–144.

Wheeler, H. (1939). *The Miracle of Life*. London: Odhams Press Limited.

Woodward, H. (1890). *Guide to the Exhibition Galleries of the Department of Geology and Palaeontology in the British Museum (Natural History)*. London: Trustees of the British Museum.

Zallinger, R. (1965). March of progress. In: *Early Man* (ed. F.C. Howell). New York: Time Life Editorial Team.

7

Fossil Collecting

There are many questions that can only be answered by palaeontology, by holding a fossil and contemplating the life it represents. There are also questions that many would prefer not to be asked, such as where we came from. But these are inevitably answered, almost as an aside to the research that human curiosity undertakes. This accidental movement across from a scientific enquiry into a philosophical one and then, worse still, a theological one, has caused considerable debate and argument. The antipathy that has been induced between science and religion has, on occasions, only been surpassed by the aggressive disdain one religion has shown for another. Like a delinquent adolescent caught out in some nefarious act, the reaction of religious groups to scientific results has been to hit out. As we shall see, much of this is due to clever men of religion with a lack of knowledge and even less understanding, trying to pass judgement on biological taxonomy of which they know nothing.

Taxonomy as understood by Aristotle (384–322 BCE) was a simple affair, and if fossil materials, most notably sea shells, were included, this did not make any significant difference to the perceived position of man in the world. This was in part because while humanity had a clear position, so did all the plants and animals, they were all individuals, so anything new or mythical could be added into the system complete and whole. There was no concept of change from one clade to another as no change was conceivable; any given plant or animal gave rise to exact copies of itself. The taxonomy that Aristotle constructed was quite simple in structure and broadly followed the theology of the time. This is based around what is described as *scala naturae,* the Great Chain of Being, with man at the top, followed by animals, plants and minerals. There was a similar hierarchy in the polytheistic religion of Greece, and later Rome, with Zeus at the pinnacle, being the King of the Gods, followed in a downwards list of lesser Gods to Man. This in itself was a reflection of a very hierarchical society, and it has had far-reaching implications for the way that both taxonomy and religion interact and all too often, conflict. Aristotle had approximately 500 animal species in his system,

Investigating Fossils: A History of Palaeontology, First Edition. Wilson J. Wall.
© 2021 John Wiley & Sons Ltd. Published 2021 by John Wiley & Sons Ltd.

which he put into order based on various factors which were simple observations of phenotype and more or less, arbitrary. The reason that the classification worked was essentially due to the teleological nature of Aristotle's ideas. As everything was in part due to a concept of final usage, it was this which guided his choice of characters with which to classify his organisms and create his taxonomy. He was careful to note that his system, while putting individual species into one of his groups, did not maintain a rigorous hierarchy as animals would have a variety of sometimes contradictory characters with which they were judged.

This taxonomy had very little to contradict it in a polytheistic system, where there was plenty of space, both theological and philosophical to fit in both the mortal and the immortal. When monotheistic religions became the social norm in the West, this situation changed. There was no longer any space for flexibility, social control became part of the regime. This included the status and position of animals and plants as well as what should be believed regarding the infallibility of the myth of creation and man's place within it.

Part of the problem with early taxonomic work was that there was a tradition of conflation of taxonomy and mythology, with an assumption that mythology was real. This, sadly, gave impetus to the idea that those with a religious background were fit to comment on the scientific subject of biology. During the Middle Ages, the *scala naturae* was developed by the Church to include all of the heavenly host. Beyond the fictional component, man was always placed at the top, not as part of the taxonomic system, but with untold arrogance, as the earthly ruler of it. By the end of the eighteenth century, the Great Chain of Being was a well-entrenched idea and although of only casual interest an example of the grip it had was that Thomas Jefferson believed in it. Such was his conviction that he denied extinction as a general concept and specifically he denied the extinction of the woolly mammoth. This, even though mammoths were only known from fossil remains. The logic of the argument was sound, but it started from fallacious assumptions. Since the extinction of the mammoth would leave a hole in nature, and nature would not allow such a thing, extinction was not possible.

There were two areas where the religious systems were going to collide head on with enlightened scientific thought; extinction and the position of Mankind in the world. Extinction had a particular taint, brought about by the insistence that God was infallible. The reasoning was that all things were created by an infallible being and an infallible being would not create a species that was so ill equipped that it could not survive. Such an animal would be less than perfect, which would be impossible for God to create, therefore extinction was impossible. This was an easy concept to encourage and enforce while fossilised material remained enigmatic in origin. There were a range of suggestions as to what they might be, none of which contradicted or interfered with the teachings of the Church. Problems started with routine discovery of fossils and the final acceptance that they represented once living things.

For sea shells, of course, it was straightforward to either claim they were the same as currently extant species, or that somewhere, in the deep and unexplored depths of the ocean they still lived. This was the situation until fossils were recognised for what they are and new and ever more devious reasoning started to be used to explain them. Indeed, the first casualty of the argument to reality was the reluctant acceptance of extinction as a real aspect of biology.

The second problem with a taxonomic approach to all living things was the position of man in evolution. This was going to produce a major clash in the nineteenth century between religion and the rational thinkers of the day. As an aspect of biology, evolution would become a considerable problem for orthodox religions from the nineteenth century onwards.

It was when fossils were first rationalised as remains of animal material, a conflict between religious expectations and rational facts became both apparent and real. One of the earliest people to wrestle with this mismatch of ideas was John Ray (1627–1705). In his *Observations Topographical Moral and Physiological* of 1673, he tried to explain the observations of fossils and the implications for the possibility of extinction (Ray 1673) (Figure 7.1). This was not just an academic exercise, Ray was trying to square the observations, which could not be gainsaid,

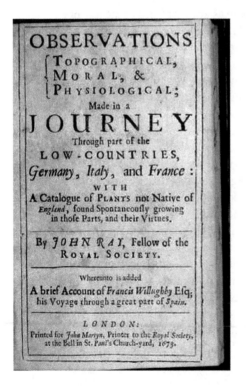

Figure 7.1 *Frontispiece of Observations Topographical Moral and Physiological by John Ray.*

with his deeply held religious convictions, which had already brought him into conflict with the authorities at Cambridge University. The way he found to get around this was for his enquiries to be limited only to those fossils which were undisputedly aquatic, more precisely, marine. This material had the advantage that it was possible to explain away the possibility of extinction. If an investigator did not want to admit extinction took place, this could be done by assuming that the species found as fossils were still alive, somewhere in the depths of the oceans. One of the examples which Ray used was of the stalked crinoids. The only stalked crinoids known at the time were fossils. Ray suggested that they were to be still be found in deep waters as yet to be discovered, since deep water exploration was impossible it was an opinion that could not be contradicted. Ray's suggestion was vindicated about 50 years after his death when stalked crinoids were discovered in deep waters of the coast of the West Indies. Unfortunately, although deep sea crinoids were found, the significant aspect is that in the class Crinoidea, all the extant species of stalked crinoids and feather stars are found in the order (sometimes, subclass) Articulata, which are distinguishable from extinct species by various anatomical features.

Even though Ray had created an explanation for some species only being known from fossils, and thereby obviating the problem of extinction invalidating the perceived perfection of creation, he seems not to have been entirely happy with the idea. This was emphasised by the ammonites. Robert Hooke (1635–1703), a contemporary of John Ray, had produced some good reasons for not just thinking that ammonites might be animal remains, but for asserting that they are. Under these circumstances, it gave Ray a different aspect of extinction to be assessed. It was noted by Ray that ammonites were something of a puzzle, primarily this was because it was such a large group of species and yet no living representatives had been found. It was the discontinuity between the possibility of extinction and consequent theological contradiction, with ammonites being animal remains that resulted in Ray being inclined to think there may be another, and inorganic, explanation for these fossil formations. This legacy of difficulty for Ray carried on in the posthumous publication *Three Physico-theological Discourses* of 1713 (Figure 7.2). What Ray managed to do was a piecemeal appraisal of fossils, so if it was less of a puzzle, why there were fossils, possibly containing indications that the fossils may be still represented as living species elsewhere in the world. This attitude towards extinction also contained the concept of *plenitude*. This theological concept did not accept gaps in nature, all species that could be present were already present, so extinction would leave a gap and in the concept of plenitude, and this was simply not possible.

This continuous attempt to reconcile extinction, and later evolution, with scriptural dogma resulted in a well-produced philosophy described as natural theology. This was occasionally also called physico-theology, which is where the title for Ray's (1713) publication came from. Natural theology can be clearly delineated

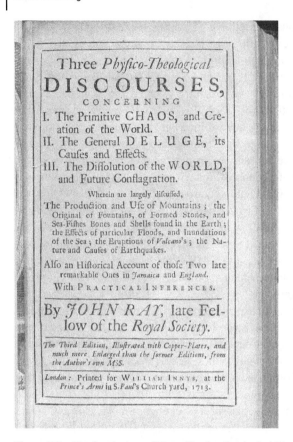

Three *Phyſico-Theological*

DISCOURSES,

CONCERNING

I. The Primitive CHAOS, and Creation of the World.
II. The General DELUGE, its Cauſes and Effects.
III. The Diſſolution of the WORLD, and Future Conflagration.

Wherein are largely diſcuſſed,

The Production and Uſe of Mountains ; the Original of Fountains, of Formed Stones, and Sea-Fiſhes Bones and Shells found in the Earth ; the Effects of particular Floods, and Inundations of the Sea ; the Eruptions of *Vulcano's* ; the Nature and Cauſes of Earthquakes.

Alſo an Hiſtorical Account of thoſe Two late remarkable Ones in *Jamaica* and *England*.

With PRACTICAL INFERENCES.

By *JOHN RAY*, late Fellow of the *Royal Society*.

The Third Edition, Illuſtrated with Copper-Plates, and much more Enlarged than the former Editions, from the Author's own MSS.

London : Printed for WILLIAM INNYS, at the *Prince's Arms* in S. *Paul's* Church yard, 1713.

Figure 7.2 *The frontispiece of Three Physico-theological Discourses, published in 1713. The complete title was quite lengthy, as was common at this time. This page faced a portrait of Ray.*

from revealed theology on the basis of where it claims its authority. Revealed theology assumes all true knowledge stems from the written word, the scriptures, as told by God or his emissary to a scribe. Natural theology, on the other hand, ignores the written scriptures and takes its proof of the existence of God from observation of the natural world and reason alone. Both of these are, of course, flawed as they start with the assumption of the existence of God and then work to prove it, usually by a circular argument. This has an exact parallel in the belief in a flat earth. The believer starts with the assumption that the world is flat, bends all data as proof (often discounting the unpalatable) and then claims the data proves the premise that the earth is flat. Neither consistent nor scientific.

Part of the interest in natural theology was as a means of joining the implied imperfection of the world by admitting extinction, to the perfection of creation.

Motivation for the creation of this philosophy came from a different arena, fuelled by the physical sciences that were becoming more amenable to mathematical analysis. Since the time of the Renaissance, in the physical sciences the elements of geocentrism and ideas of a homocentric universe had been whittled away. These ideas were being gradually replaced by physical observation backed up by a mathematical analysis of them, resulting in hypotheses that gave reliable predictions. Such an approach had quickly changed our knowledge of the orbit of the planets and the place of Earth in the solar system. It had also been made plain by the dominant religions that this was an unacceptable trespassing on their area of teaching. The conflict between church and the newly evolving physical sciences boiled over into some very public disputes, as with Galileo. In such one-sided arguments, the individual generally fared badly. However, the delight of science is that it works – whether you believe in it or not and so the reductionist ideas which started with the physical sciences were becoming tools for natural history, specifically the nascent study of biology, including Man.

Given this background, natural theology became a very useful way of deflecting the argument that studies of science in general and biology in particular would lead everyone into atheism. In a broad sense this philosophy was thought to be a way of learning about God by studying the natural world. Into this arena both Robert Boyle (1627–1691) and John Ray had things to say as advocates of natural theology. Boyle had a very mechanistic attitude to the natural world. Later centuries would regard it as a reductionist view, where a series of simple rules resulted in a complex machine. While Boyle and Ray were approaching the attempt to reconcile Church and science from a practical observer's viewpoint, another original thinker, Rene Descartes, was looking at the same problem from a different angle. Rene Descartes (1596–1650) came to a reductionist view of the world, while still requiring a Creator, via philosophical considerations alone. Descartes believed the essence of natural science was to be found by discovering relationships which could be mathematically described. By inference of this, if it is true, then all natural science could be unified by mathematics and all natural explanations could be described by mathematics.

Descartes started from the only certain knowledge which he had, which was that he was a thinking being. This starting point was based on a line of reasoning which, briefly summed up, was that he doubted everything, including his doubts, therefore the process of doubting was a process of thinking. Moving from that premise, he claimed that a perfect being could not originate from the thoughts of an imperfect being, like himself, so there must really be a perfect being who originated the idea. This process of starting from an origin point and developing all thoughts from there was summed up in the aphorism, cogito ergo sum, I think therefore I exist. Descartes went on to the conclusion that all things can be explained in mechanistic terms except soul and free will. This would be the most extreme reductionist view of the world while still requiring a Creator.

Although as a complete philosophical enterprise the set of Cartesian ideas of the world can be disputed, it was, and still is, highly influential. This mechanistic approach was a considerable problem for religious groups as it would suggest that God was not a direct creator, but a somewhat lesser lawgiver, setting out the rules from which the physical world developed once set in motion. Robert Boyle circumvented the contradiction by suggesting that although the world could be understood as a mechanistic structure, it is far too complicated to have formed in any other way than by an intelligent designer, in his theosophy, God. In this way he managed to have a mechanical explanation and an omnipotent God, as he expounded in his work of 1675, *Some Physico-theological Considerations about the Possibility of the Resurrection* (Boyle 1675). Boyle went into considerable detail regarding the functional nature of resurrection, wherein we can take any substance suitable for a body and reinstate a soul, thereby making the same person again. Besides the rather massive jumps in reasoning the book was reviewed in *Philosophical Transactions of the Royal Society*, 31 December 1674. This was date was written $167^4/_5$ and represents a social change which caused some confusion at the time. This year date was marked down in this way because in Europe the original Julian calendar had been replaced by the Gregorian calendar, which was much more accurate. It had the effect of disconnecting the 2 calendars by 10 days. What was even more confusing, the older Julian calendar had the start of the year on 25 March, but with the new, Gregorian calendar, the year start was 1 January. By giving a joint date the Philosophical Transactions was showing that it was both modern and international, one or other of the dates being correct wherever you were reading it. I was not until 1752 that the United Kingdom and colonies adopted the Gregorian calendar, by which time the drift was such that 3 September was followed by 14 September.

During this period, Robert Hooke (1635–1703) was also looking closely at the fossil record. Best known for his microscopy, he put this to good use in *Micrographia* (Hooke 1665) by comparing petrified wood with new wood. He is credited as being the first person to look at fossils using a microscope and went on to look at many more fossils which he compared with extant species. These were published sometime after he had died in *Posthumous Works, Containing the Cutlerian Lectures* (Hooke 1705). While Ray was considering the possibility that ammonites were not organic in nature, Hooke took a different line of reasoning which circumvented that problem. He assumed that ammonites were the remnant shells of once living species and therefore because none were known to exist they must have become extinct. The next step in the logical process was more controversial for Ray, as it included another point which went against traditional religious views of speciation. Using a line of reasoning based on a teleological belief of creative direction, it was possible for Hooke to accept extinction as long as there was a concomitant possibility of new species coming into existence. By this combination of extinction and speciation, this would maintain the situation of plenitude in the natural world, the fullness of nature. Hooke considered it possible that these new species originated to fill the gaps left by extinction in a manner analogous to breeding new varieties of domestic animal. The arguments made by Hooke were not seen as persuasive by his contemporaries and were largely ignored. Trying to gain acceptance for the composite idea that fossils were extinct forms and new ones could arise was really just too radical for the educated of the day, all of whom would have been taught religion as an essential part of their education. A classical education would have been essential for any career and would never have involved attempts to radically re-work traditional interpretations of theological scripts. It would require a very independent mind to think through the observations and realise that a new explanation would be needed to make sense of the world. Although Hooke's acceptance of extinction and the concomitant creation of new species was in later years taken as a demonstration of early Darwinian thinking, this is unlikely. It was in all respects a philosophical attempt to marry theological doctrine with observable phenomena. There was no coherent mechanism suggested that would have moved his ideas from descriptive form to explanatory. It is far more reasonable to suggest that this was explaining an observation, extinction, and then simply adding in speciation to maintain completeness in the natural order.

The work of Boyle and Descartes which used a mechanistic approach to living things was put into an analogy by William Paley (1743–1805) that would become better known than the author and his works. Paley was Archdeacon of Carlisle and in 1802 published *Natural Theology, or Evidences of the Existence and Attributes of The Deity* (Figure 7.3). This contained what has become a very well-known analogy of a watch and a watchmaker. He suggested that if a watch was found, its intricate construction would immediately be understood to be the result of an intelligence, the watchmaker. Going further, the watch was too complicated to have been constructed by chance. As there were so many natural structures of greater complexity than a watch, there must be an even greater intellect that designed and constructed the universe. This spurious argument neglects the fact that all the components of an inert watch are required for it to do a single task, while the random development of functions gives biological systems flexibility in what it does and what it may do. This analogy was given a wider airing when it was used in the title of Richard Dawkins' book, *The Blind Watchmaker* (Dawkins 1986). The work by Paley is of importance primarily because it revived a largely retreating idea. By the end of the eighteenth century in Europe, natural theology was under siege, while in Great Britain, the idea still retained some credibility. This was presumably due to the hold the Church had on University departments and education in general.

Paley made extensive use of examples from earlier writers, such as Ray, who were clear in their commitment to natural theology. Although published in the

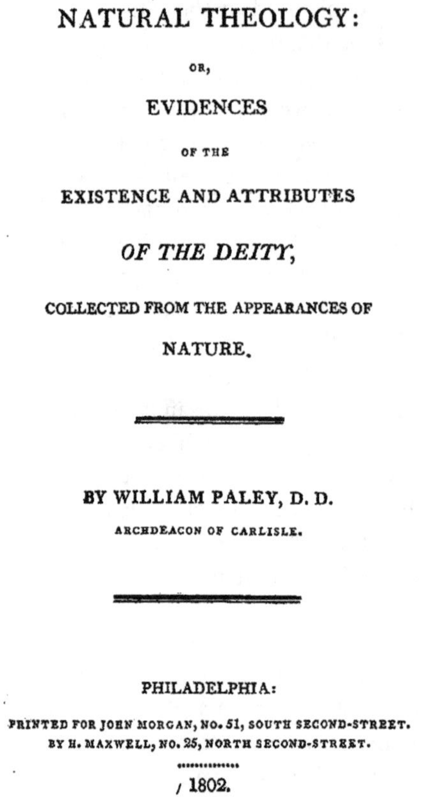

NATURAL THEOLOGY:

OR,

EVIDENCES

OF THE

EXISTENCE AND ATTRIBUTES

OF THE DEITY,

COLLECTED FROM THE APPEARANCES OF

NATURE.

━━━━━

BY WILLIAM PALEY, D. D.

ARCHDEACON OF CARLISLE.

━━━━━

PHILADELPHIA:

PRINTED FOR JOHN MORGAN, NO. 51, SOUTH SECOND-STREET.
BY H. MAXWELL, NO. 25, NORTH SECOND-STREET.
••••••••••••••
⁄ 1802.

Figure 7.3 *Frontispiece of* Natural Theology or Evidences of the Existence and Attributes of The Deity *by William* Paley (1802).

nineteenth century, it was apparent that *Natural Theology, or Evidences of the Existence and Attributes of The Deity* was looking backwards to the previous century where it was not the age of the world that was at issue, so much as the Lamarckian ideas of change. Paley was much more comfortable with the concept of a static world with an unchanging biota than with any form of change, either extinction or speciation.

By the time the nineteenth century was underway, powering forwards towards industrialisation and the perceived man's mastery of nature, ideas were changing regarding the age of the world. With fossil finds becoming ever more numerous, natural theology was becoming more and more difficult to defend. It has, however, survived in part with the development of creationism as a series of wild and inconsistent explanations of the observed world. Part, if not the whole problem, was that ideas were entrenched as a form of social control. It may not have appeared to be like that at the time, but like so many processes, the subconscious requirement is to maintain the *status quo* and like all rules for that purpose, it was regarded as essential that no flexibility was allowed. This made the acceptance of accumulating evidence difficult to ignore, but if it could not be ignored it could be denounced. Such a process requires no evidence, it is enough that it contravenes accepted policy. If outright denouncement does not work, then bend the evidence so that it bolsters the accepted view. This was the situation with natural theology. The slow process of understanding and acceptance of fossil evidence for the processes of both geology and biology took a long time, in fact, throughout most of the nineteenth century. It was not simply on religious grounds that ideas of evolution and extinction, as an intrinsic part of the processes of biology, were causing trouble. The extreme range of time that was required was troublesome in parts of the scientific community as well. This disquiet with extreme geological ages was regardless of Darwin and his evolutionary ideas being accepted in broad terms. Although Darwin had been very circumspect in his treatment of humanity, it was going to be the inclusion of *Homo* in zoological phylogenies and taxonomies that would give credence to perpetuating creationism well into following centuries. With increasing religious dissent amongst the educated middle class, it was a fear among the conservative clergy that not only was their religion being forced to modernised, but it was being fundamentally undermined.

The shift of philosophical query was away from the simple problem of extinction, implying a fallibility of design, to something more fundamental. As the nineteenth century progressed, religious ideas shifted ground, turning the ideas of religion towards the idea of man at the top of the phylogenetic tree. This accommodation of scientific ideas had an unforeseen result as it squeezed out the possibility of man as a fallen species requiring salvation. Man became a product of the family tree, whether static or dynamic. Even if this could be used to demonstrate the guiding power of God, it did not allow for a direct hand of creation.

The argument between humanist and religious groups of all sorts, this was not confined to Christian groups alone, came from a position of intolerance. Although this position started as a dislike of questions that could undermine scripture, it appears to have developed into a fear of change. The possibility cannot be ignored that this antipathy was a problem because change would reduce the power of the clergy. These arguments primarily took place, or at least started, in the first half of the nineteenth century. By the time that *Origins* was published in 1859, another change was developing which seems to have been fired by misunderstanding. Even if there had been a clear idea in the minds of the protagonists of exactly what the arguments and discontinuities in agreement were, there would have still been intractable debate between those believing that the translated scriptures were from God and those that relied on evidence. One side took the evidence and made deductions, the other side started from an assumption and fitted those parts of the evidence they wanted into their thought system. This misunderstanding of evidence was going to spill over and become the lynch pin for modern creationism, where facts are chosen that fit an argument, rather than taking all verifiable observations into account.

In 1860, seven months after *Origins* had been published, there was a debate at a British Association for the Advancement of Science (BAAS) meeting in Oxford. This was to become well known in later years, even though it is very difficult to verify precisely what was said and to whom. Finding what was said in this debate is problematic, mainly because there is no verbatim report of the meeting and most of what is known comes from written material produced many years later (England 2017). Even so, it is primarily remembered for a confrontation between Samuel Wilberforce, the Bishop of Oxford and Thomas Henry Huxley, a friend of Darwin. Also known to be present among others were Joseph Hooker, who was not only a close friend of Darwin, but also a botanist and explorer. Robert Fitzroy, who was the Captain of the Beagle and a skilled meteorologist, was also present. The presence of Fitzroy at the meeting was coincidental, he was in Oxford to present a paper on meteorology, a subject for which he was justly famous. It was reported that he was in the audience and stood up brandishing a Bible while entreating the audience to believe in God, not man. However, the anecdote which is well known from this meeting was between Huxley and Wilberforce. Huxley was a staunch supporter of Darwin and quite prepared to speak publicly in support of the broad ideas contained in *Origins*. This is in contrast with Darwin, who did not like public meetings, especially those that might become argumentative or contentious, as a consequence of which, he was not even present at the BAAS meeting. Huxley was a professional scientist, one of the new breed of investigators who made their living from scientific endeavours. He recognised that whether he agreed entirely with the content or not, Darwin's ideas opened up an entire avenue of investigation. His understanding of this was because while previous commentators had accepted that the fossil record indicated change in the natural world, no plausible mechanism had previously been formed. With a potential explanation for a process that had always relied upon the hand of a creator, this became amenable to scientific investigation.

At the BAAS meeting of 1860, the clash took place which would become a part of scientific folklore. This unusual position of a scientific meeting in the middle of the nineteenth century attaining an almost mythical status strangely rests on an anecdote, the only written evidence of which comes from a printed account in a weekly publication from London called *The Press* (Jensen 1988). The exchange took place between Bishop Wilberforce of Oxford and Thomas Henry Huxley. Wilberforce had been previously described by Benjamin Disraeli as 'unctuous, oleaginous, saponaceous', which gave rise to his nickname of 'Soapy Sam'. Wilberforce asked Huxley whether it was through his grandfather or grandmother that he claimed decent from a monkey? This deliberate insult was countered by Huxley with the longer riposte that he would not be ashamed to have a monkey for his ancestor, but he would be ashamed to be connected with a man who used his great gifts to obscure the truth. The comments by Wilberforce were both offensive and embarrassing, but on the day it was Joseph Hooker whose rhetoric carried the argument for the Darwinian camp. The difficulty that Wilberforce faced in the debate seemed to have stemmed from ignorance and misunderstanding of the ideas contained within *Origins* (Darwin 1859). This contrasted with a similar debate that had taken place in 1847 where Wilberforce had floored Robert Chambers over his book *Vestiges* (Chambers 1844), published in 1844, which had come in for a great deal of criticism from both the clerical and scientific communities.

While being an ardent supporter of Darwin, Huxley was not entirely convinced by the arguments that he produced in favour of speciation. Huxley thought that a domestic breeding programme could be used to clarify the situation. Huxley was less convinced by gradualism as suggested by Darwin than the saltation of large-scale spontaneous changes and mutations. The biggest arguments taking place during the middle of the nineteenth century were mainly associated with the question of humanity with regard to the natural world. It seemed to most of society that the position of Man was at the top, in charge of the Earth. It did not seem reasonable that humanity was in the same position as the rest of the natural world, it was perceived that in some way people were different. This has clouded our attitude to such an extent that resource exploitation has been carried out with scant thought for the resulting damage to the environment.

In the 1860s, there was still a residual dislike of the methodological means by which Darwinian evolution could take place. But at the same time, Darwin's basic concept of evolution was perhaps surprisingly, quickly accepted in scientific circles. Problems with accepting evolution were not so much the idea as the process. Although it was in 1866 that the experimental work of Mendel was first published, this was in a journal of a natural history society (1866) and was largely ignored. What seemed to slow its dissemination was the complex mix of where the research was carried out, where it was published and within the paper, no hint of its wider implications. This changed when it was translated by Bateson and re-published in the Journal of the Royal Horticultural Society (1901). Understandably, there was still a major conceptual gap between the nascent science of genetics and any thought of applying this knowledge to evolution. What had been published by Mendel was a description of small-scale variation, suitable to explain variation in plant hybrids, as the title suggested, but there was no hint of it being a mechanism for bigger changes, as in speciation. What it did show was that it may be possible to discern a method whereby evolution as it was understood could take place. This left evolution in a position of having a functional description, but still with no mechanism with which to explain the large-scale changes that speciation needed.

While disputes over the mechanisms of evolution were being actively debated, there was a question that was being avoided through lack of data and that was what is the precise position of man with regard to evolution. For many it was an unanswerable question, and for some it was a question that should not be asked as it was obvious that humanity was outside the common range of science. During the first half of the nineteenth century, there was a plausible argument made for this idea, which would also take Man out of the normal run of taxonomy and phylogeny. This was the dearth of fossils which could in anyway demonstrate that mankind was just another species. There were very few fossils which could be shown to be human or which came from a human precursor.

Lacking hard evidence for the position of man in the natural world, it was reasonable that as the dominating species, man should be at the top of the tree. This was, in broad terms where the notion of man as the top predator/organism originated. It was all perfectly reasonable to the religious believers, given that man had *dominion over the fish of the sea, and over the fowl of the air, and over the cattle, and over all the earth, and over every creeping thing that creepeth upon the earth.* (Genesis 1:26), apparently, although we have made a pretty poor job of looking after it. Such an argument for man's superiority also went well with the scientific community that was struggling with Natural Theology and trying to reconcile the observed world and the scriptural world. If this was an idea to be accommodated, then it would be necessary to create a hierarchy of life; some method of describing the relationships within and between species was needed that fitted the evidence. For palaeontologists this was not a particular problem as it was understood that the deeper the fossils, the older they were and so it was implied that species appearing low down in the strata were not just older, but earlier forms. The leap that was taken then was to cause considerable misunderstanding for a very long time. The older forms were seen as earlier and therefore more *primitive*. It was the use of 'primitive' which was the problem. In the way it was used, it referred to an earlier form, not as is frequently implied in modern usage, inferior or at least simpler. However, taking shape in taxonomy and phylogenetics was a developing use of hierarchies to denote biological importance. Concepts of 'primitive' and 'advanced' within biological thought could be seen, clearly represented in some of the contemporary works. One such was *Foot-Prints of The Creator, or, The Asterolepis of Stromness* (Miller 1849), originally published by Miller in 1849. In this work he describes a process of 'degradation' where the original form effectively de-evolves to a more primitive state. The example he chooses for this is the loss of legs in snakes. Miller was a very competent palaeontologist, but his inherent teleology shines out in his descriptions of ill-omened birth, for the origin of snakes. He considers that 'the degradation of the Ophidians consists in the absence of limbs'. More was to come:

> I am also disposed to regard the poison-bag of the venomous snakes as a mark of degradation; – it seems, judging from analogy, to be a protective provision

of a low character, exemplified chiefly in the invertebrate families, – ants, centipedes, and mosquitos, – spiders, wasps and scorpions. The higher carnivora are, we find, furnished with unpoisoned weapons, which, like those of a civilized man, are sufficiently effective,... *(Hugh Miller Foot-Prints of The Creator, or, The Asterolepis of Stromness Eleventh edition 1869)*

It is the assumption that a civilised man would not use a poisoned weapon that underlines his attitude of superiority towards the natural world and belief in a bidirectional evolution. As an idea that is quite reasonable, the snakes had, after all, lost their legs, but that is not the same as degradation, but adaption to an environment in a different way.

The shift of emphasis during the second half of the nineteenth century had moved from unexplained catastrophe to acceptance of gradual change as a process. This helped in development of the idea of a bottom and top in development and of a species position in the world. As Haeckel put it:

It is evidently perfectly absurd to assume a distinct new creation of the whole world as definite epochs, without the crust of the earth itself experiencing any considerable general revolution. *(Haeckel 1876)*

There was, however, a distinct element of teleology still present as few scientists could distance themselves from their early education, which would almost invariably have had a large religious content. The most influential manifestation of this came with the way that relationships were laid out for the perusal of the public. Although we are used to seeing genealogical charts, starting with the proband at the top and a cascade downwards of progeny through succeeding generations, this is a more recent development of the mediaeval Tree of Jesse (Figure 7.4). This depiction of the lineage of Christ through his ancestors back to Jesse of Bethlehem is inverted, with Jesse at the bottom and Christ at the top. It was developed from a line in Isaiah, the oldest known version being from the eleventh century. The significance is the genealogy moving from the bottom upwards, precisely as evolutionary depictions also run, towards the top of the tree in the shape of mankind. In one particular way this is counter-intuitive as the prevailing phrase makes use of the word *descent*. This was found in the theory of descent, as used by Lamarck as early as 1809, and in *The Descent of Man* (Darwin 1871).

This particular orientation, with the most ancient at the bottom, is a logical one for a geological system, as the oldest strata are the deepest. By starting at the bottom it reflects geological reality, as it is most often seen, remembering that folding and tilting can alter this simple image. Onto this image of strata Haeckel then took a step further and mapped onto it a suggested phylogeny of vertebrates. This first appeared in 1866 (Haeckel 1866) and then in a modified form in his 1876 publication,

Figure 7.4 A thirteenth century depiction of the Tree of Jesse from the *Scherenberg Psalter of c 1260*. Original at the Badische Landesbiblioteck, Kalsruhe, Germany. *Source:* https://commons.wikimedia.org/wiki/File:Cod_St_Peter_perg_139_Scherenberg-Psalter_7v.jpg#/media/File:Cod_St_Peter_perg_139_Scherenberg-Psalter_7v.jpg, Public Domain.

The History of Creation. The most significant aspect of these is that by overlaying palaeontological data on geological data, there was an implicit age becoming associated with the various species, both living and dead. This was the first time that the two had been so closely associated. Of course, there was an incomplete fossil record for many groups, but with the new found knowledge that clades give rise to new species it was easy to effectively ignore gaps in the fossil record. Such gaps were no longer assumed to be reflections of the true state of the fossil record, but simply gaps as yet unfilled. This was a significant challenge for those religious groups that were clinging to spontaneous creation by a deity; the gaps had been seen as precisely that, species having been created without a visible lineage. These illustrations of the development of life, with increasing numbers of species as time progressed, were an essential part of Haeckel's support for Darwin. Even though they were described as a phylogeny of vertebrates, they left out specific references to mankind on the chart.

Haeckel was in many ways outspoken in his written works, and this came in the form of fulsome support for the ideas of evolution, originally without detailing

Man as part of the equation. After the publication of *The Descent of Man* in 1871, it became possible to include man as a biological entity, responsive to the same forces of evolution as any other organism. This was still, however, with one unstated proviso; that modern man was different. It was alright to speculate on the origins of humanity, as long as it was accepted that modern man was the pinnacle of the evolutionary process. This implied a large belief in a teleological world, a directed evolution culminating in man, which helped soften the blow for a retreating religious absolutism.

The ease with which man could be taken on as part of nature was partly because the story, as it was shown in the fossil record and in ontological investigations, was observational. The past that was being described was effectively inviolate, including the apparent arrival of man on the scene. By putting humanity at the top of the story, it became a crowning glory of the natural process. How humanity got there was open to interpretation, so religions of all types could claim that the hand of god directed evolution, or simply laid down the rules by which it took place. With the exception of the religious extremes, evolution became quietly, a more acceptable idea amongst the educated.

Into this arena Haeckel produced *The Evolution of Man* in 1897. In Volume II of *The evolution of Man* Haeckel represents the Pedigree of Mammals (Table XXIV) and Pedigree of Apes (Table XXV) (Figure 7.5) as straightforward linear phylogenies, both with man at the top. Although this could be argued to be an affectation of expectation, it was simply following a conventional system of pictorial display. Bound between these two tables there was another illustration Pedigree of Man (Plate XV) (Figure 6.3) which was quite different and set a precedent for later depictions of man's position in the natural order. While in the preceding and following tables Haeckel had no indication of a temporal; element, this was implied in the plate. The depiction was of a tree with single-celled organisms at the bottom, moving through perceived levels of complexity to Man at the very top.

Although illustrations of the Jesse Tree had been made using a tree in a literal way in the past, Figure 6.3 was the first time a tree had been used to illustrate what became known as the tree of life. This image was a powerful way to illustrate many aspects of the natural world. It had an implicit time line associated with it, and a clear depiction of primitive life at the bottom in some way evolving into more complex forms. It also made sure that there was an understanding that man was at the top, the ultimate aim and result of evolution, directed in some way either by a creative hand or rules laid down by a creative hand.

At the time that Haeckel was publishing, the fossil record of humanity was very limited, this would only develop during the twentieth century into a useable palaeontological story. Nonetheless, the image of the tree of life was of such influence on thinking that it persisted well into the period when fossil finds were changing attitudes to the position of humanity in the natural world. These depictions of a progressive tree aiming towards Man were often associated with small

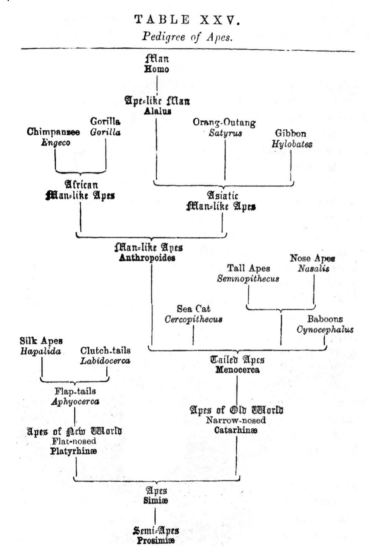

Figure 7.5 The pedigree of apes, as depicted by Haeckel (1897). By putting Man clearly at the top he recognises Man's position as a part of nature, but simultaneous implying his dominion.

vignette images of a stylised person at the top of the illustration. All other life was somehow beneath the crowning glory of evolution (Brightwell 1939; Mayo 1944). Needless to say that while we might be a more recent arrival as a species than many others, this does not mean we are in any way a more highly adapted species than any other. We are, of course, fantastically successful, but we may be self-limiting in our success, depending upon whether we survive the negative environmental changes that we have created for ourselves. Putting the emphasis on Man as the top species in illustrations of phylogenies and taxonomies based on temporal knowledge has created some assumptions which are difficult to justify. One of these is the Drake Equation.

Searching for extraterrestrial life has been going on since man first looked to the heavens, but it was only in the second half of the twentieth century that an attempt was made to formalise the process, and quantify the expectations. This took the form of an equation created by Frank Drake, a radio astronomer, in 1961. This was a probabilistic calculation which tried to estimate the number of habitable planets there were, and how many of those would evolve intelligent life. This is where the prior assumptions engendered by images of man as the pinnacle of evolution make itself felt. Having been steeped in the concept of man as on top by virtue of brain size and intelligence, it was never questioned in the Drake equation and rarely elsewhere. The problem is simply put; man is a product of evolution and as such intelligence is simply another characteristic, ecosystems do not need an intelligent member any more than they need any other particular species trait. After all, the planet has only had an intelligent species for a very small part of its existence and great swathes of full and balanced ecosystems, separated by geography and time, survived and became climax systems without an intelligent head. As an aside to this the Drake Equation has been described as a piece of scientific flummery, a set of guessed variables multiplied together to give no particular answer.

One of the first human fossils that was recognised came from Belgium and was found by Phillip-Charles Schmerling in 1829. The skull is now known to be Neanderthal, Schmerling being the first to recognise the prehistoric nature of the material and the associated implications. It was originally thought to be a modern human skull and to Huxley, as modern, it would have been of no consequence for the purposes of taxonomy. Even when Haeckel was writing about the origins of Man in 1897, fossils did not play a significant part in his deliberations. By the time the twentieth century arrived, it had become increasingly difficult for mainstream religions to engage in the scientific debate about human origins because it was no longer an area in which they had anything to contribute.

During the late nineteenth and early twentieth century, while there was both an increase in fossil material found for both humans and other animals, there was also a retrenching of religious attitudes. The element of conflict arose when attempts are made to interpret religious matters to accommodate the scientific data. This has two fundamental problems, the first is that there is no reason for scientists to be involved in any discussion, because there is no scientific aspect which requires debating, so any debate is purely religious, *ipso facto* it is for the churches and religions to sort out their problems and not to try and deny observable fact. The second problem that such debates have is that they start from a basic premise that an ancient text is true, rather than the expression of a curious mind trying to make sense of a complicated world. This assumption of truth involves all manner of attempts to define palaeontological finds and data as a test, a coincidence, a strange quirk or just not relevant. Many of the interpretations discussed by theologians start from a framework interpretation of Genesis, assigning various time scales and periods to the stated days of creation. Although many groups have tried to look at palaeontology and evolution as part of the natural order, all deistic theologies start from an immovable assumption of a creator, with no justification and no evidence at all.

For many religions it was thought that while evolution may have some merit for describing adaptions of the body, it did not help describe creation. There was still a belief, stated or tacit, that whatever was happening in the biological world, man was the exception to all of it. It was in this climate that Gregor Mendel, as a cleric, produced his seminal work in 1866 on genetics. This was later given a wider audience in 1901 by Bateson. At the time it was clear that these new ideas of inheritance and genetics had no implications for evolution. This was just as well as during the second half of the nineteenth century the Catholic church of Rome set down a series of significant proclamations which impinged on evolutionary theory.

Pope Pius IX was elected in 1846 and would rule until 1878, so he spanned the period of publication of *Origins* in 1859. During the First Vatican Council of 1869–1870, Pius IX declared the idea of Papal Infallibility and set up a system of rules which can only be described as setting the Roman Church against evolution and cogent interpretation of fossils. The general idea was summed up in the suggestion that anyone who denied the creator or asserts there is nothing but matter were anathema. In this context anathema means the individual is excommunicated. This was of considerable influence, but was more a demonstration of social control than reasoned argument. A century later it was still being debated by theologians as to the precise meaning of the declaration while also trying to modify the outcome. For about one hundred years, until the papacy of Pius XII in 1950, the whole question of evolution was broadly ignored by the Catholic church. In 1950 the Vatican announced that the Church does not forbid research and discussion on evolution of the human body. By being explicit about the corporeal, this still tacitly blocked discussion of the development of human morality, which Darwin had spent considerable time writing about in *The Descent of Man* (1871). The problem that all religious coteries have is the unprovable idea of a Creator. This conundrum was approached in a different way by Bertrand Russell (1872–1970) by not trying to refute the initial argument, but by changing the burden of proof. In this he used an analogy now referred to as Russell's Teapot. The analogy was used to demonstrate that the burden of proof for an argument lies upon the person making unfalsifiable claims, rather than shifting the burden of disproof to others. The basic argument is to suggest that there is a teapot orbiting between earth and Mars – on the assumption that nobody can prove that it is not. This is, however, so unlikely that it is just not considered worthy of being taken into account in general arguments. He goes onto say 'I think the Christian God just as unlikely'. It can be seen from this that any theological debate regarding evolution and the fossil record in verifying it is essentially a simple religious discussion held between the simple religious, it is not in any way a scientific discussion.

As the protestant attitude moulded to an acceptance of evolution and the changes, both moral and physical, associated with the origin of mankind in the nineteenth and early twentieth centuries, political and economic problems were going to change many religious attitudes. One of the changes that occurred was a resurgence in a fundamentalist attitude to scriptures. In 1925 Tennessee passed the Butler Act, introduced by John Butler, who was a state representative and previously head of the World Christian Fundamentalists Association, not withstanding its name, a peculiarly US society. The Butler act specifically prohibited the teachers in public schools in the USA from denying the Biblical account of creation. This effectively stopped the teaching of evolution. Later in the same year, 1925, both Mississippi and Arkansas passed similar laws through their legislature. It was only in 1968 that the Supreme Court of the USA struck down these laws, not because of the irrational nature of them, but because they seemed to be violating a constitutional point that no laws should influence the regulation of religion. Very soon after Tennessee had passed the Butler act, it was put into use.

There is no doubt that the facts of palaeontology and scientific investigation have caused some alarm in religious circles. This often manifests itself in some form of legislative lashing out, but the best thing about science is that it does not matter what you believe, it works anyway.

References

Bateson, W. (1901). Experiments in plant hybridization. *Journal of the Royal Horticultural Society* 26: 1–32.

Boyle, R. (1675). *Some Physico-theological Considerations about the Possibility of the Resurrection*. London: H. Herringman. Reviewed- *Philosophical Transactions of the Royal Society* 9, 111, 31 December 1674.

Brightwell, L.R. (1939). What evolution means. In: *The Miracle of Life* (ed. H. Wheeler), 9–24. London: Odhams Press.

Chambers, R. (1844). *Vestiges of the Natural History of Creation*. London: Churchill.

Darwin, C. (1859). *On the Origins of Species by Means of Natural Selection: Or the Preservation of Favoured Races in the Struggle for Life*. London: John Murray.

Darwin, C. (1871). *The Descent of Man*. London: John Murray.

Dawkins, R. (1986). *The Blind Watchmaker*. London and New York: Norton and Company.

England, R. (2017). Censoring Huxley and Wilberforce: a new source for the meeting that the *Athenaeum* 'wisely softened down'. *Notes and Records (Royal Society)* 71 (4): 371–384.

Haeckel, E. (1866). *Generelle Morphologie der Organismen*. Berlin.

Haeckel, E. (1876). *The History of Creation*, vol. 2 (trans. E. Ray Lankester). London: Henry S. King & Co.

Haeckel, E. (1897). *The Evolution of Man*, vol. 2 (trans. E. Ray Lankester). New York: D. Appleton and Company.

Hooke, R. (1665). *Micrographia*. London: Martyn and Allestry.

Hooke, R. (1705). *Posthumous Works, Containing the Cutlerian Lectures, and Other Discourse*. London: Richard Waller (for the Royal Society).

Jensen, J.V. (1988). Return to the Wilberforce-Huxley debate. *British Journal for the History of Science* 21: 161–179.

Mayo, E. (1944). *The Story of Living Things and Their Evolution*. London: The Waverley Book Co.

Mendel, G. (1866). Versuche über Pflanzen-Hybriden. *Verh Naturforsch Ver Brünn* 4: 3–47.

Miller, H. (1849). *Footprints of the Creator*. Edinburgh: William Nimmo. 11th Edition 1869.

Paley, W. (1802). *Natural Theology, or Evidences of the Existence and Attributes of the Deity*. (American Edition). Philadelphia: John Morgan.

Ray, J. (1673). *Observations Topographical Moral and Physiological Made in a Journey Through Part of the Low Countries*. St Paul's, London: John Martyn, Printer to The Royal Society.

Ray, J. (1713). *Three Physico-Theological Discourses. I the Primitive Chaos, II the General Deluge, III the Dissolution of the World*. London: William Innys.

8

The Rise of Fossil Fraud and Special Sites

Almost for as long as fossils were recognised as being rare and unusual, even before their origins were questioned, frauds and hoaxes have been recorded. Broadly these have taken two forms, one rather more serious for palaeontology than the other. The first is selling fakes and the second is passing fakes off as real, or claiming a real fossil was found at an inappropriate site. The least of the problems comes from the sale of fake fossils. These are usually sold as souvenirs, and as long as the professional realises what they are and the amateur does not mind, little if any harm will have been done. The real problems start when a deliberate deception is perpetrated, and this can have repercussions on careers and the understanding of the earliest forms of life on Earth.

Although it may not be immediately apparent, the problem of fakes and frauds can be compounded by the problems of high prices paid for genuine fossils. The high value of fossils has also been driven by illegal collection and sale of smuggled fossils. It has always been a delicate balance between private collectors for whom the financial value of a fossil is on a par with the scientific value and the, usually academic, collector for whom the fossil itself is the primary interest and the financial value of little or no interest. As has been said, 'demonising commercial trade is a lazy way of simplifying ethical issues in palaeontology' (Liston 2016). With commercial collectors needing to make a living and for the most part behaving in a responsible manner, it is worth remembering that many of the scientifically important fossils that have been found over the years would have been lost to erosion without them. On the other side of the argument, local laws have to be adhered to and an ethical respect for the origins of fossils always considered. A delicate balance does have to be achieved which is sometimes missed by the imposition of laws that were formulated for another purpose but encompass fossil material by way of accident. An example of this is the *Kulturschutzgesetz* of Germany. This is a Culture Protection Act which was designed for, and mainly affects, the art world but does seem to impinge on some

Investigating Fossils: A History of Palaeontology, First Edition. Wilson J. Wall.
© 2021 John Wiley & Sons Ltd. Published 2021 by John Wiley & Sons Ltd.

museum activity (Kreye 2016; Daumann 2017). The almost accidental inclusion of fossil material in legislation stems from the perception of an apparently seamless join between archaeology and palaeontology. This will have the inevitable result of poor regulation, where regulation is required, of the sale and movement of fossil mateial. Black market trade in fossils does have a long history and sometimes this feeds the market for fakes and frauds. Where a fossil is known to have been illegally procured, close examination of its provenance may be overlooked, so a clever copy will serve a similar purpose, by avoiding close scrutiny itself. The fake originating in this way may be enhanced to increase its value in a market where rarity and aesthetics rather than science drive the system.

There is a case where a very unusual motivation seems to have driven the production of a hoax, if indeed, that is what it was rather than a genuine belief in the beast. This is the naming of the mythical Loch Ness Monster. Although searches of the Loch had been carried out at various times over a long period, it has never been demonstrated that such an animal exists. However, the motivation for this particular search and naming activity seemed to stem from the best of intentions. If the Loch Ness Monster existed it would, without doubt be an endangered species, it may be an individual animal, or at best a very small population. To gain protection under UK law, an endangered species has to be named, so it was necessary to name the animal even without proof of its existence. Based on a blurred underwater photograph purportedly showing part of the body and a fin, it was described by Scott and Rhines as *Nessiteras rhombopteryx,* (Scott and Rhines 1975). This name was later noted by *The Daily Telegraph* to be an anagram of 'monster hoax by Sir Peter S'. The reason for including this here is twofold, and the first is to demonstrate that a hoax can be motivated by very good intentions. The second is that while even the existence of it as an animal is disputed, if it did exist, one of the suggestions given credence is that it is, or was, a relict plesiosaur. Other suggestions as to what it might be have included a long necked newt and a large bristleworm.

Perhaps surprisingly, there is a situation in which legal traffic in palaeontological material increases the illegal traffic in modern material. This paradoxical situation has developed with the trade in mammoth ivory. With increasingly warm summers in the Siberian tundra, the export of ivory from mammoths has become a significant business. This is legally exported to countries in Asia where it is carved into decorative jewellery and figurines. The traffic is legal simply because this 'ice ivory' comes from extinct species, so they are no longer endangered species. Most of the trade originates from *Mammuthus primigenius*, as this is the species which consistently produces high-quality carvable ivory. This trade is booming, according to CITES trade monitoring there was approximately 17.3 tonnes traded in 1997, up to 95 tonnes by 2012 and by 2018, well over 100 tonnes. As would be expected with an increasing trade, alternative supplies are always being looked for. In this case, the easiest supply which can be tapped in to is illegal ivory from poached elephants (Flannagan 2019). Consequently, legal trade in mammoth ivory is being used to launder illegal trade in elephant ivory by mixing the two sources together and claiming it is all ice ivory. Wildlife monitoring has demonstrated that deliberate mislabelling has taken place by dealers in China, Hong Kong, Burma and the USA, turning illegal elephant ivory into legal mammoth ivory. The problem for all concerned is that it is not easy to determine the origin of the ivory simply by looking at it. There have been social changes wrought by this trade as well as fuel to poaching. One of the areas of in which trade in ice ivory has significantly developed is a region of northern Russia called Yakutia (Figure 8.1). This area is now the Republic of Sakha and includes islands of the East Siberian Sea which produce large amounts of ice ivory. In this whole area, there was a long tradition of local farmers supplementing their income by digging out mammoth tusks for sale. With a steadily increasing demand for ice ivory, the local population has felt they are becoming increasingly sidelined as commercial mammoth hunters are moving in to take part in the multimillion dollar trade. In an attempt to curtail the unwanted spin-off trade in elephant ivory, CITES, the Convention on International Trade in Endangered Species, suggested in 2016 that there should be a curb on trade in ice ivory. To even attempt control of an animal product, it is necessary to have a named species from which the material can be identified. It was suggested that the species for which this would have been done was *M. primigenius,* the species whose ivory was most often used for carving. Although CITES can prohibit trade in extinct

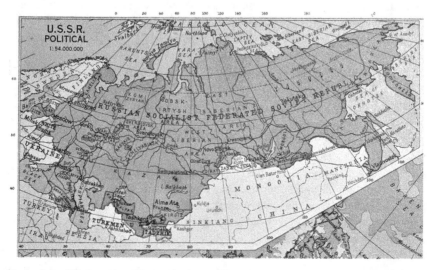

Figure 8.1   *The Republic of Sakha, marked on this map of 1942 as Yakutsk, a constituent part of the Soviet Union, and large-scale producer of mammoth ivory.*

species when it resembles a species which is living but endangered, this is a tricky procedure and one more influenced by politics than common sense or science. For this reason, the amendment was withdrawn from the CITES meeting in Sri Lanka in May 2019. It is not a technical issue to identify the species origin of these tusks it seems to be a political belief that trade is more important than any other human activity and certainly more important than an animal species, living and dead.

One of the aspects of a science such as palaeontology that deals in such exotic species that they almost defy the imagination is that they can stray into fiction. This is neither fraud nor fake as the intention is to entertain rather than mislead. By way of demonstration of the scientific sense of humour, even the Bulletin of the British Ecological Society has been involved (Hogarth 1976). This article was so well crafted that it was reprinted many years later in 1989, in the Journal of Biological Education (Hogarth 1989). In this article, it is suggested that dragons were common in the eighteenth century, but the fact that no remains are known was due to lack of care on the part of the general population. These mythical animals arose as a distinct group, apparently about 5000 years ago, with many types being found, flying, marine and polycephalic.

Palaeontology is not the only science that has been trialled by frauds, fakes and mistakes, from alchemy to the claims of cold fusion in physics. Mistakes and dishonest production of results can have long-term repercussions, so it is incumbent on scientists to spot the problems and put them right. This is the hope of peer review and the scientific method, although as we will see, this can sometimes fail. There is no doubt that in palaeontology there is a significant commercial pressure that has fuelled prices, tempting manufacturers of fakes and dishonest traders in legitimate fossils. The financial value of fossils has in itself been affected by the increasing interest in palaeontology that has been seen in the general public in the second half of the twentieth century. This growing interest and excitement surrounding all fossils received a boost with the publication of *Jurassic Park* by Michael Crichton (1990) and the subsequent film of the same name. Whether such publicity was a significant influence on the outcome of the auction of an almost complete skeleton of *Tyranosaurus rex*, it is hard to know. On that particular occasion, the mortal remains, nick-named Sue, fetched $8.4 million. This was paid for by a group in which the Field Museum of Chicago was the central player. These sums are enormous for what is, after all, a biological specimen, even though collectors of orchids and butterflies have paid vast sums for single rarities in the past, never amounts on the scale of an almost complete dinosaur skeleton.

There is no doubt that rarity value comes into the equation when considering cost and value and this goes back at least as far as Mary Anning in Lyme Regis. Anning, along with many other collectors, made money by selling small fossils to trippers and holidaying visitors that had been collected from the coastal cliffs of Dorset. Although the sums were small, Mary Anning sold the first Ichthyosaur she found for £23 (Figure 8.2). This high price was an indication of an interest that

Figure 8.2 *The fossil Ichthyosaur originally discovered in 1811 by Mary Anning in the cliffs of Lyme Regis. This was originally sold for £23 to Henry Henley of Norfolk.*

would steadily develop over the coming years and significantly alter prices. Even though it was a consortium headed by a museum which bought Sue the *T. rex* for many millions of dollars, it is unusual for an academic institution to have the resources to fund such purchases. For private collectors of immense wealth, of which there are many, the inflation of prices being paid for unusual or unique specimens does tend to put them in direct competition with scientific institutions. With this growing commercial interest in fossils, it was inevitable that, like paintings by famous artists, it would attract the attention of forgers and producers of fake fossils specifically to fool the purchaser and make more money than a simple, known, copy would.

While in later years, the issue of fakes and forgeries has generally been about money, this has not always been so. Some of the best known fossil frauds have been less about money than reputation and these are recorded because they can blight careers and even put reputable journals to the test of their honesty. One of the earliest cases of forgery that is on record was motivated by entirely different thoughts than financial gain. Johan Beringer (1667–1740) was Court Physician at Würzburg in Germany, where he was also Dean of the Faculty of Medicine at the University. In the 1720s, Johan Bartholomew Adam Beringer was routinely making trips to investigate the quarries at Mount Eibelstadt, close to Würzburg in Germany. As a keen geologist, he routinely collected fossil sea shells and ammonites and as a curious man he wondered about their origin. Like all educated people of the time, his formative years would have been spent learning scriptural stories of creation and the Flood. He came to believe that these shells were in some way associated with the Deluge, as biblically described. While Beringer was trying to fit his observations with his religious convictions, he became the butt of an extremely unkind hoax. This hoax, as far as we can tell, was motivated by professional rivalry. Whatever the reasons and motivations, it took some considerable time to execute, and when it was done, the story to conclude unhappily. Beringer had been investigating the quarries at Mount Eibelstadt for some time when in 1725 he was presented with some unusual fossils. These had been planted at a site that Beringer routinely investigated and seemed to him to represent various animals. These included lizards, crustaceans and frogs among the many different forms that were supposedly represented. Although his original finds of marine shells and ammonites were genuine, the later and more complex species were fakes. These had originally been carved by one of his assistants, Christian Zanger. Zanger, it is thought, was not entirely self-motivated in his deception as he was paid to do this by Professor Ignatz Roderick, Professor of Geography, Algebra and Analysis, also at Würzburg University. Another colleague was also involved, Privy Councillor and University Librarian, Johann Georg von Eckhart. While it seems Roderick paid Zanger, Eckhart helped in the deceit by completing the reproduction 'fossils'. Calling the carvings fossils is deceptive since they do not appear as any fossil seen before or since (Figure 8.3).

It was because of the enigmatic nature of fossils during the eighteenth century, both in origin and taphonomy, that these unlikely representations were taken seriously by

(a)   (b)

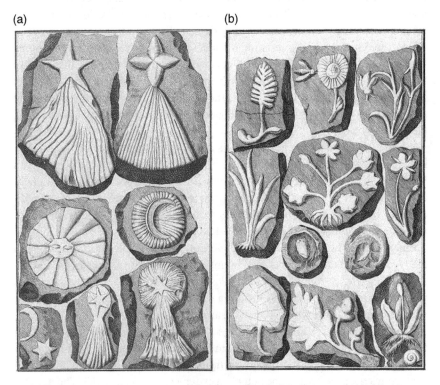

Figure 8.3 (a and b) *Two examples of carvings used to fool Beringer, as depicted in his work,* Lithographiae Wirceburgensis, *published in 1726. As can be seen, they are poor carvings of a range of subjects.*

Beringer. They were given credibility by both their geology and geography, being found in a quarry well known to yield marine fossils. He took these carved stones very seriously and described Mount Eibelstadt as a horn of plenty, where Nature had deposited what would normally be scattered amongst many different sites. The fake material was cut from limestone and some even had visible chisel marks, but these were explained away by Beringer. It was a year after his first finds, in 1725, that Beringer published *Lithographiae Wirceburgensis* (Beringer 1726), a meticulous description of the found stones. It is from the included lithographic representations of them that we know so much about the range and nature of the fraud. In this work, he described the fossils with the tools marks as being the work of a very meticulous sculptor, which he rationalised as the marks left by a chisel wielded by the hand of God.

By this time, both Roderick and Eckhart had suggested publicly that the fossils were fakes, but it did not shake the conviction of Beringer that these were divine products. As the deception proceeded it was obvious that it was getting out of hand, so the hoaxers started to leave more and more extreme material with 'clear images of the sun and moon...'. It became ever more radical with the final pieces of stone carved with inscriptions, supposedly of the name of Jehovah in Latin, Greek and Hebrew. Even so, Beringer accepted the material at face value and since they were inscribed, he assumed they must have had a different origin to the more normal fossilised material which he came across in the quarries he visited. To this end, he developed a hypothesis that the inscriptions were in some way associated with a memory held within light that could be imprinted onto softer material, such as mud or clay. With the rumours of a hoax being spread by Roderick and Eckhart, reaching the ears of Beringer, it became apparent that he would have to take some action against his two colleagues. The way he did this was to take them to court. Part of the court proceedings are available, and it is apparent that the two hoaxers were unmasked. It also came out that they wanted to damage and discredit Beringer for personal reasons in that they considered him arrogant. The unhappy outcome of this was that Beringer, having been taken in so thoroughly by what were seen with hindsight as obvious fakes, lost his reputation. At the same time, Both Eckhart and Roderick lost their university positions, in fact Roderick had to leave Würzburg entirely to gain employment. Some of the stones have survived the intervening years and are held at the Oxford University Museum in the UK and Teylers Museum, Haarlem, in The Netherlands. These stones are now referred to as *Lügensteine* – lying stones.

Palaeontology is obviously not exempt from accidental misinterpretation of results, but just like any discipline, malice is very difficult to guard against. The inflexible attitude of Beringer made him especially susceptible to a hoax which played upon his beliefs. The entrenched ideas that he held were of such strength that they were able to rationalise the discordant nature of some of the found stones masquerading as fossils. This determination to persist with a view that is gradually being whittled away is part of science, waiting until the tipping point comes and there is too much contradictory evidence to persist with the original hypothesis. The reason that Beringer is still so well known as having been hoaxed is that he would not give up his assertion that the fake fossils were actually real. So dogmatic was he that it resulted in his action in court. Other hoaxes and fakes that have been documented have not generally resulted in legal action, although even mistakes and misinterpretations can cause problems.

When the first fossils of *Ichthyosaurus* were discovered, form had to follow the available evidence. So although a dolphin-like profile was apparent, there was no evidence for how the tail or unboned fins were constructed. Early images of *Ichthyosaurus* were consequently given a more reptilian tail and only fins for which there was an associated bone structure, so there was no dorsal fin (Figures 8.4 and 8.5). It was not until 1891 that evidence was available from high-resolution fossils found at Würtemberg that there was a dorsal fin present. These same fossils corrected the structure of the tail fluke (Figure 8.6).

During the nineteenth century, there was a period which has picked up the emotive epithet of the 'bone wars'. This took place in the USA where the problem of a clash of personalities became far more acrimonious due to a simple mistake. Although this did not involve, so far as is known, either a fake or a forgery, it does give some indication of the motivations and high stakes that were running in palaeontology at that time. It is also an example of how personal animosity can affect an entire science and the public perception of it. It is worth looking at the competitive nature of this story in some detail as it is lead by personalities rather than active scientific disputes based on interpretation of factual evidence. The two

Figure 8.4 *Model* Ichthyosaurus *from Crystal Palace, London. Originally commissioned in 1852, the reptile is shown with no dorsal fin and a reptilian tail. The apparently amphibious position was as much for display as a reference to life style.*

PLATE II.

Ichthyosaurus communis.
Length about 22 feet.

FISH-LIZARDS.

Fishes, *Dapedius, etc.*

Ichthyosaurus tenuirostis.
A small species.

Figure 8.5 *Illustration of two species of* Ichthyosaur *from* Extinct Monsters, (Hutchinson 1892), *without dorsal fins.*

central protagonists were Edward Drinker Cope (1840–1897) and Othniel Charles Marsh (1831–1899). Unusually, these two scientists were both financially independent, Marsh from a $100 000 legacy from his uncle, George Peabody, and Cope from his family shipping business set up by his grandfather. Although they were from different backgrounds, they started out quite amicably, even though there was a fundamental difference in their scientific approach to the interpretation of fossil material.

Edward Cope was a supporter of a broadly Lamarckian interpretation of evolution, while Othniel Marsh favoured a more straightforward interpretation of evolution by natural selection, as propounded by Charles Darwin. They had met in Berlin over a period of days in 1864, although this may have been the start of their personality clash that ultimately cost them both a huge amount of money and social acceptance. Initially, there seems to have been a certain amount of scientific professionalism, with Cope naming a fossil species after Marsh; *Colosteus marshii*, while Marsh named a species of mosasaur after Cope; *Mosasaurus copeanus*. This latter is cod Latin and could be taken as a thinly veiled insult.

The animosity grew with stories of Marsh visiting a marl pit in New Jersey that was part of an investigation being carried out by Cope. This had already been a source of scientifically valuable fossils, including *Hadrosaurus,* but this did not

APPENDIX V.

ICHTHYOSAURS.

IT is unfortunate that news of the highly interesting discovery at Würtemberg came too late for our artist to make a new drawing to

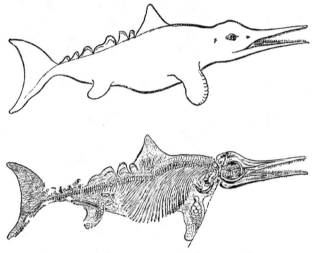

FIG. 38.—*Ichthyosaurus tenuirostris,* from Würtemberg.

show the dorsal fin and large tail-fin, etc., described by Dr. Fraas ;[1] but, by the courtesy of the proprietors of *Natural Science,* we are

[1] Ueber einen neuen Fund von *Ichthyosaurus* in Würtemberg. *Neues Jahrbuch f. Mineralogie,* 1892, vol. ii. pp. 87–90. The same author has published a valuable monograph, with beautiful plates, entitled *Die Ichthysaurier der Süddentschen Trias- und Jura-Ablagerungen.* 4to. Tübingen, 1891.

Figure 8.6 *Illustration from the Appendix V. of* Extinct Monsters, *(Hutchinson 1892), where new information showed both a dorsal fin and a dolphin-like tail fluke.*

stop the encroachment on this apparently personal fiefdom by Marsh. It was said that he had bribed the quarry men to send him any future fossils of note, rather than passing them on to Cope. This marked the start of a deterioration in the relationship between the two palaeontologists which would become progressively worse as insult and perceived slights spilled from the professional into the personal. The animosity began to show in scientific attacks on each other in published papers. The point at which it became no longer possible to resurrect the professional relationship, if not the personal one, is generally traced to a single event that took place in about 1870. Cope had described a new marine reptile, *Elasmosaurus,* which he had suggested had a very oddly constructed backbone. This description had been published in 1868, but it was not until 1870 that Marsh came to view the fossil in person, at which point he claimed that Cope had put the head on the wrong end, which would account for the unusual backbone (Figure 4.21). In making this observation, Marsh was quite correct, but there had already been many reproductions of the plesiosaur *Elasmosaurus* with an incorrectly positioned head. Understandably, Cope was greatly distressed by the potential of the error to tarnish his reputation as a palaeontologist. It should be said that with the head on the right end, as an animal it does seem to be of an improbable construction. Cope argued his case until it was decided between Cope and Marsh that Joseph Leidy should arbitrate as to the most likely and scientifically defensible picture. The humiliation of Cope over this matter must have been compounded by the correction being published sometime later by Joseph Leidy (1870), a palaeontologist who had taught Cope comparative anatomy many years earlier. Cope had already published his account of the fossil (Cope 1869a, b), so when it became obvious that the mistake would not be ignored, it is said that Cope tried to buy back all the printed copies of his paper that contained the mistake. By this time, however, the original illustration had been copied and reproduced and had even been translated into a complete scene of various fossil reptiles, although some of these were so generally inaccurate that it made little difference to the scientific reputation of Cope. About 20 years later, Marsh published his own account which described the mistake of Cope in putting the head on the wrong end of *Elasmosaurus,* including illustrations of it (Marsh 1890). The very next day Cope published a reply in the same newspaper, admitting and accepting the points that Marsh had made, but also pointing out that he had corrected the error many years previously and that it was no longer of any interest.

By the time that the furore over the incorrectly positioned head had more or less been settled, vast fossil beds had been discovered in the interior of the USA. These were being investigated by Marsh and Cope, along with other groups, one of which was run by Joseph Leidy. By this time, Cope was a staff member of the USA Geological Survey, run at the time by Ferdinand Hayden. It was Hayden who regularly sent fossil material to Leidy for study, but this was to change with Cope joining the Geological Survey. This situation was made more difficult when Cope went to the Eocene fossil beds of Wyoming to collect fossils at the same time that Leidy was there with the same intention. Hayden did try to smooth over the problem, but was essentially powerless to control Cope as although nominally employed by him, he was paid no salary and therefore regarded himself as independent.

The competition between Cope and Marsh was far more than a simple dispute over a single fossil, these two were in dispute over reputations. Reputations based entirely on palaeontological evidence for their own interpretation of the fossil record. It even extended as far as an expedition where Cope, either deliberately or inadvertently, employed diggers who were already supposed to be employed and paid for by Marsh. Such events did nothing to ease the distrust between the two men, and so, by 1873 no vestige of professional acceptance, one for the other, remained. Even so, Marsh, Cope and the older Leidy were all working in the same geological formations of the central USA. General animosity between Cope and Marsh extended to published material where synonyms were commonplace and taxonomic conflict became normal. This stemmed from both collectors being so dogmatic in their position that they would be aware that a species they had excavated had already been found and described, but they would feign ignorance and rename the specimen as if it was an original find. This inevitably caused some confusion among other palaeontologists as names and synonyms had to be sorted through and reorganised to a more stable system that could be relied upon.

The personal fallout for the two protagonists was considerable (Jaffe 2001), and the high cost of supporting field workers significantly reduced Cope financially and Marsh was so unreliable at paying his field workers that he alienated some of his most useful and productive assistants. Still the animosity continued, with Cope producing a notebook which he had kept over many years, of the errors and perceived wrongdoings of Marsh. When this notebook was published in serial form, in the New York Herald, it had a consequence which had been completely unforeseen. This unforeseen consequence was of alienating many of the scientists that had been associated with either Marsh or Cope over the years. Both of the entrenched men, Cope and Marsh, ended up with little of their respective fortunes left by the time the feud effectively ended with the death of Cope in 1897.

There is no doubt that the result of this unfortunate and long-term animosity between the two palaeontologists had contradictory outcomes. The first is the waste, lost opportunities and indiscriminate damage to sites in a headlong rush to extract the biggest and best fossils at any cost. The second plays on this same idea. If it were not for these two scientists working against each other, many of the finds and discoveries may have waited decades to be made, some even never found at all. It has been suggested that this competition, which certainly got out of hand very early on, was detrimental to the reputation of American palaeontology for many years if not decades. This may be true in some quarters, but in *Extinct Monsters*, published in 1892 (Hutchinson 1892), the work of Cope and Marsh is extensively quoted, without any hint of dissent between the two men, or suggestion of questionable results. This unhealthy competition between the two men and their field collectors was fuelled by personal feelings but could not have been carried out without their personal fortunes, much of which was squandered in the process. Throughout the period of nearly 30 years that this was going on, the naming and deliberate renaming of species that had already been described inevitably caused difficulties for other workers in the field. These competitive attempts to be the first to describe and name species were often not mistakes, but deliberate. The rush into publication of every new fossil species also saw errors enter into the record beyond just names. One such more complicated mix-up included Marsh putting the wrong skull on a skeleton and declaring it a new genus. Even so, new species were described from the fossil beds of the USA, including *Stegosaurus* (1877) and *Allosaurus* (1877), described by Marsh. While not directly associated with fakes and frauds, the example of Cope and Marsh gives a good idea of the high stakes and motivations that have been found in palaeontology. These had developed ever since it was realised that fossils meant something more than just strange formations and could make a reputation. There is also the intrinsic curiosity of discovering the new and putting your stamp on it. These items were from a past which is, to all intense and purposes, as alien as life on another planet, in this case separated by time rather than distance.

When the subject of a hoax becomes larger than was anticipated and bigger than can easily be controlled, no end of problems arise. The problem with these pieces of deception is that however they start out, if they are taken seriously by palaeontologists they have the potential to ruin careers. One of the greatest acts of fraud perpetrated on the scientific community became so well known that the name still lingers in collective memory of the general public. This was Piltdown Man, a hoax on a grand scale that for a while, when it was believed to be true, skewed the work of many palaeontologists and archaeologists who were convinced that this human relic was real.

The story of Piltdown Man starts in 1912, when Charles Dawson (1864–1916) claimed he had found a fossil skull, later described by Dawson and Woodward (1913) and Woodward (1913). Dawson trained as a solicitor and pursued a hobby as an archaeologist and palaeontologist. He discovered many various fossil remains while he lived in the south of England, some of which were new to science and were named after him as acknowledgement. There were many finds of various sorts, a large number of which were investigated after the death of Dawson, and found to be fakes themselves. Although it was initially publicised in 1912, the original skull was said to have been discovered in a Pleistocene gravel pit in 1908. Dawson contacted Arthur Smith Woodward, who was Keeper of Geology at the British Museum (Natural History). It was Woodward who then reconstructed the bone fragments into a skull and produced the initial publication describing the find (Dawson and Woodward 1913) (Figure 8.7).

After the initial discovery, Woodward, along with Dawson, discovered more fragments, including a jaw bone and some primitive tools made of flint and some of bone. Woodward postulated an age for the skull of about 500 000 years and named it *Eoanthropus dawsoni*. The additional finds were all reported as having come from spoil heaps associated with the gravel pit that had yielded the original skull fragments. The discontinuity between the skull and the jaw was an early source of contention. While Woodward maintained that Piltdown man was a missing link between apes and man, others put a more modern interpretation on the relics.

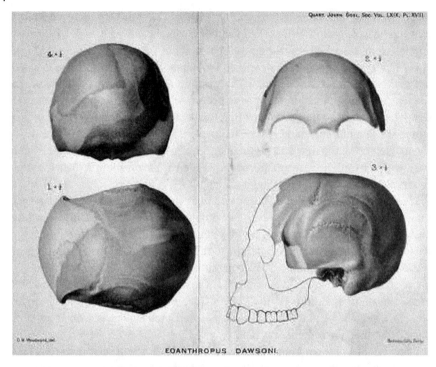

Figure 8.7 *The illustrated reconstruction of the skull of Piltdown man. This illustration, from* Dawson and Woodward (1913), *was one of several fold-out illustrations in this paper.* Source: Dawson and Woodward (1913).

Arthur Keith, another distinguished anatomist, while believing in the authenticity of the remains, was an advocate of them being from of a much more recent species, far more in line with modern man. His reconstruction differed from that of Woodward and received the Latin binomial *Homo piltdownensis.* Although there was much debate about its age, there was general acceptance among palaeontologists that Piltdown man was a true and genuine relic of humanity.

Debate regarding the nature of the Piltdown material seems to have started in 1913, when David Waterston analysed the radiograms that had been made of the jaw bone (Waterston 1913). These had been published in the British Journal of Dental Science in October 1913 and Waterston had published a letter in *Nature* in November of the same year, where he suggested that the jaw was from a chimpanzee and had been inappropriately attached to a human skull. During the following years, with an additional skull fragment found somewhere in Kent by Dawson, he never divulged where it was supposedly found, Piltdown man was taken as genuine. In *The Miracle of Life* (Wheeler 1939), a description of the broad understanding of the status of Piltdown man was laid out. This included the difficulty of explaining the apparent vast geographical distance between hominid fossils of the

Far East and Western Europe. To account for this, a suggestion was made of there being a line of immigration from the East in a westerly direction, this would assume an eastern origin for humanity. In comparing Piltdown with Peking man, Wheeler comments that it was an outcome 'had Nature been experimenting along now one, now another, line of advancement towards modern man'. This was a teleological slip which was quite common and his entire work is redolent with the old view that evolution was in some way working towards modern man. His statement of the apparent discontinuity between the two ancient skulls, one from China and one from Europe, was borne out in the comparison, which broadly came down to Piltdown having an advanced skull and primitive jaw and Peking having a primitive skull and more advanced jaw.

The story with Piltdown man was causing a problem by creating a human lineage that did not have a discontinuity in it brought about by the seeming modern cranium and pre-human jaw. That it had no apparent chin was the problem, the chin being a human trait. Not only was the jaw the cause of the problem, but it was the jaw which started to set the story straight. An anatomist was looking at the teeth and specifically the wear patterns. He noted that the type and pattern of wear looked as though they had been worn by an abrasive, rather than the action of chewing. This immediately set in motion a number of researchers to look again in detail at the original finds. The outcome of this was that when details were published in late 1953, it was said that the skull was mediaeval, the jaw was from an orangutan, with various diagnostic details broken or filed off and the teeth were from a chimpanzee.

It was an extraordinarily disruptive announcement to studies of hominid evolution and caused considerable problems for the genuine researchers who were working on the subject. Even for those not directly involved in such work, the belief that these fossils were true and reliable, however inconvenient, was essential, and to find that they were fraudulent raised many hackles. One for whom this deceit was particularly onerous was A. S. Romer. In his publication of 1933, *Man and The Vertebrates*, Romer gives several pages over to a discussion of Piltdown Man as a genuine fossil of considerable importance to studies of human antiquity. Understandably, a detailed description of the skull and jaw were given as at this time it was a true fossil from the Pleistocene which had something to say about the origins of humanity. It is quite easy to look at the text with hindsight and imagine that Romer was questioning the validity of the finds by comparing the skull with a modern *Homo sapiens* and the jaw with higher apes. Indeed, he does suggest that if the skull had been found under different circumstances, it would be taken as from a modern human. Even though it was recognised that the jaw may not have come from that actual skull, because they were found in close proximity, the balance tipped in favour of them being matched. Trying to fit the Piltdown fossil into a human phylogeny was going to be difficult. One of the most innovative suggestions was that there may have been two lines of descent in the

Pleistocene. The first would have run as far as Neanderthal man and then died out, and the second was suggested to have started as Piltdown man, which gained an anthropic jaw and from there developed into modern man.

Later editions of *Man and The Vertebrates* were renamed *The Vertebrate Story* and Romer revised the editions in line with increasing knowledge. With the discovery of the Piltdown, fraud Romer changed the section heading from 'Piltdown Man' to 'Piltdown Hoax'. His understandable dislike of the hoax can be summed up in a quotation from a later edition of *The Vertebrate Story* (Romer 1959).

> Scientists may differ widely (and often violently) as to conclusions to be drawn from their work, but the basic principles of the scientist's code is absolute honesty in presentation of their facts. And because of their own high standards of truth, the scientist is liable to be taken in when he comes in contact with an amateur lacking the same code of ethics. *(Romer 1959)*

We expect the same standard of honesty from our lawyers, sadly in this case Dawson was not only an amateur scientist but a professional lawyer as well. The quote from Romer sums up the problem with this sort of fraud. Fossil material tends to be rare and fragmented, and it is not easily amenable to repetition, in this case finding more material of a corroboratory type. In the physical sciences, repetition is straightforward, with other people simply repeating the same experiment. So when such material as this is found, it is assumed that what the discoverer says regarding the material is true, both in time and place. Consequently, a carefully engineered fraud that is deliberately designed to deceive is not always easy to uncover. As a result, when the motivation seems to be little to do with money and more to do with reputation, it makes it all the more difficult to uncover the truth.

Part of the infamy associated with the Piltdown hoax originates with the widespread dissemination of the case, the importance of the original find had it been genuine, and the importance of it being found out as a fraud. This fame resulted in the epithet 'Piltdown' becoming synonymous with fossil fraud. Such was this effect, that other cases, such as the fake fly in amber and Archaeoraptor, were described in the popular press as 'The Piltdown Fly and the 'Piltdown Chicken', respectively. These two cases are described later.

There is one well-documented case where a simple motivation of professional advancement seems to have got out of hand and resulted in a massive fraud. This particular fraud had a significant additional aspect in that it involved some well-known additional players who were unwittingly drawn into the story. The central character is V.J.Gupta, a palaeontologist in what was the newly formed Department of Geology, Panjab University. The development of the reputation of Gupta hinged around his work with fossils which he claimed were from the Himalayas. Initially, he was happy to claim the finds, but senior scientists were encouraged by him to put their name to his work as a sign of validation. The subject of these ranged from Graptolites which were not previously found in the Himalayas (Sahni and Gupta 1964; Berry and Gupta 1967), and Devonian fishes, also previously unknown in the area (Gupta and Denison 1966). He also published papers on Conodonts that were claimed to have been found in the Himalayas. The list was long and a whole career was based on them, but they were eventually found to be false discoveries. When the various reviews of the case were completed and the enquiries reported, it was seen that the falsification of data came in three different forms (Shah 2013). The first was attributing samples from one area to another. These reassignments were often from completely outside the area. Some of the fossils had even been bought from commercial dealers. The second area of fraud involved using the same material, not just fossils, but the same photographs as well, in several papers where the claimed site of discovery was different in each paper. Thirdly, it was found that he had used other peoples' illustrations, by re-photographing them and then claiming that they were of his own specimens. The discovery of this fraud set back the work of many people who had assumed the publications to be genuine and had incorporated the data into their own work. Inevitably, Gupta came out of the review of the fraud very poorly, retiring from the University in 2002. Such a case as this has a clear motivation associated with it, but sometimes, by the time that a fraud has been discovered, the perpetrators are imposible to track down and we can only speculate about the original motivation.

We may never know exactly where or when a notorious fossil fly originated, but it does seem likely that it was associated with a highly skilled individual who was constructing a small item simply as a saleable 'fly in amber'. The story starts with the acquisition of a fly in Baltic amber by the British Museum (Natural History), now the Natural History Museum. It was bought from the estate of a distinguished German Entomologist, Friedrich Loew. It was first mentioned by Loew in his personal catalogue in the 1850s, but it was not until 1922 that the Museum bought the specimen along with about 300 other pieces of amber from the collection of Loew. It lays in the collection as an unremarked specimen until 1966 when Willi Hennig made a detailed study of it and recognised the unique nature of the encapsulated fly. The dipteran fly in question was *Fannia scalaris* and by being found in amber which could be given an age, it put the origin of this group of insects as much earlier than had been considered likely. It demonstrated that this particular group of advanced flies was extant at least 38 million years earlier than had previously been thought. This shifting of dates was something of a problem, but as it was an almost perfect specimen, there was no disputing what it was. As a consequence of this, it became one of the greatest of the museum's entomological treasures. It was a case, very similar in many ways to the Piltdown hoax, where the assumption of honesty by the receiving scientists, caused so many problems. The difference was that there was no known perpetrator.

Except for a happy accident, this anomalous fossil would have continued to vex palaeontologists indefinitely. In 1993, Andy Ross, a researcher at the Museum, was investigating the sample using a microscope with an incandescent light, this was the normal light source at the time, before light-emitting diodes became commonplace. As indicated in the name, these lights use a lot of energy to generate light, most of which is dissipated as heat. In this case, heat was the important part of the equation, probably in conjunction with the age of the sample. The sample cracked, but it was not a natural flaw, and the nascent fracture propagated along an otherwise invisible line. It turned out that the manufacturer of this erroneous fossil sample had cut a piece of amber down the middle, carved a small space, put the fly in and then resealed the block with resin (Ross 1993; Grimaldi et al. 1994). It is generally considered that this was not a deliberate attempt to pervert evolutionary theory, but a simple way of making money by selling interesting specimens to collectors. Nonetheless, for such a small deceit, it certainly had considerable repercussions over the years. Production of a specimen such as this would have most likely been carried out by a jeweller familiar with the process of forming gemstone doublets. Producing a doublet is the process whereby precious stones are attached to a backing of lower value material to increase the overall value. It is most often used with opal, but in these cases for a very practical as well as financial reason. If the opal is of a particularly good quality, but too delicate to be used in jewellery, it can be bonded to a supporting material, such as black onyx or black opal to give it strength. Occasionally, to give support from both sides, an opal may become a sandwich with a clear stone on top, in which case it is a triplet. As this is a skilled activity, it seems certain that to make an undetected amber composite would have been the province of a skilled craftsman.

An equally spectacular fraud to the fly in amber described above involved a vertebrate fossil, found in 1997, which could have had even greater repercussions within palaeontology and our understanding of evolution. The reason it did not become problematic was that it was discovered to be a fake before it became a mainstream problem. It was primarily the way in which announcements were made regarding this fossil, before it was fully investigated, that brought it resolutely into the public eye. In 1997, a fossil was found at Chaoyang in Northeast China of a feathered and toothed animal with the tail of a reptile. This was sold to a dealer and smuggled to the USA in 1998. Even though this specimen had only just been brought into the public domain, it had violated two local statutes. The first was it being dug up and sold on by a private individual and the second was it having been sent out of the country without permission. We know that it was dug up by a farmer and although strictly against the rules, this is a common event in China and Mongolia, where farmers in areas of high concentrations of fossils supplement their income by being part time fossil hunters and dealers.

What the farmer was said to have done was accidentally break the fossil. Being fully aware that a complete feathered dinosaur was of much greater value than a partial fossil, the finder located nearby, feathered legs and tail. From this a jigsaw, a chimaera, was created. When it had been sold on and had been smuggled out of China, it arrived in the USA, where it was investigated with a view to publishing a complete description and official name. It was about this time that commercial pressure became involved in the scientific process. While trying, unsuccessfully, to get a peer-reviewed article into *Nature* or *Science,* it was agreed that *National Geographic* magazine could effectively launch this very important fossil, which was still regarded as genuine, at a press conference to coincide with a magazine article written by C.P. Sloan (1999).

Although there were many reasons for the scientific papers to be rejected, one of them was the dawning knowledge that the fossil had been illegally removed from China and should be repatriated there. It is quite normal for scientific journals to reject palaeontology articles where fossils cannot be demonstrated to have been collected and transported legally. The National Geographic press conference went ahead with the fossil being summarily given the name *Archaeoraptor liaoningensis.*

Closer investigation of the fossil by several palaeontologists raised concern regarding the validity of the specimen as a single and complete fossil. One of the investigating palaeontologists, Phil Currie, even noted that the 2 feet were actually from slab and counter slab. It was also seen on close examination that there was no connection between the tail and the body. Altogether it was thought that the putative *A. liaoningensis* fossil was made up of fossils from five different origins, but most likely from the same area (Dalton 2000). These details were passed on to the owner, along with the suggestion that it could be a fraudulent fossil was the question of legality. Throughout this period, it was regularly pointed out that regardless of the nature of the fossil, it had been illegally removed from China. Shortly afterwards, in 2000, the fossil was repatriated to China.

This is particularly interesting because although the original deceit was carried out on the basis of improving the value of a found object for the finder or dealer, the similarity of the situation with the fossil fly in amber ends at that point. Both seem to have started out as a simple attempt to maximise profit by augmenting the appearance and perceived value to a collector. The fly in amber then accidentally became part of a palaeontological fraud. This happened by the specimen passing into the hands of a zoologically aware collector, rather than a collector of *object d'art.* What happened after *A. liaoningensis* was originally sold, was rather different as this was always going to be an object of study, rather than just decoration. It was this aspect which rapidly brought to light the inconsistencies in the fossil, but it was the commercial imperative of the fossil which made it so infamous in so short period of time. Without the involvement of a commercial enterprise, willing to set up a press conference designed to publicise the fossil via the *National Geographic* magazine and thereby increase sales, the whole business would have had far less impact in palaeontological circles. It may even have passed unnoticed by the popular press and avoided the unhappy description of 'Piltdown chicken'. The continued commercial pressure seemed to have pushed the fossil forward against a growing tide of knowledge that it was not as stated, and it had been illegally removed from China. This fits in with the Kuhnian idea of scientific progress by punctuated equilibrium (Kuhn 1996). When the accumulated evidence could no longer be gainsaid and the reality became visible, repatriation of the fossil was inevitable. It is in many ways surprising that the Chinese should want such a fraud back, but this was probably more to do with the principle – it was illegally removed and therefore should be returned – rather than any actual loss of scientific material.

The question of fossil ownership is a perplexing one in many ways, but not a question that should be ignored or shied away from. There are many ethical questions that have to be addressed, some of which are very difficult, especially when the supposed fossil turns out not to be a reliable fossil, or even not a fossil at all. Because the trade in fossils is so lucrative, forgeries such as *Archaeoraptor* are not so unusual as palaeontologists would like. Sometimes, the skill of the forger is of such a level that it takes an expert analysis to recognise that the fossil has been manipulated and is made up of different specimens from different areas.

It is no surprise that given the way that fossils have been manipulated for profit, both financial and professional, any truly unusual fossil will be viewed with understandable suspicion. If along with that is included the problem of the purported fossil having been smuggled out of the country of origin, without full documentation of where it originated, and it can be seen how some important material may be lost. There is an interesting example of this confusion associated with a single fossilised specimen which took some considerable time to unravel. The specimen involved was named *Halszkaraptor escuilliei* and was named in recognition of François Escuillié a fossil trader who also helped to repatriate all manner of fossils back to their country of origin (Cau et al. 2017). When it first came to light, this particular fossil had already been smuggled out of Mongolia, probably by a circuitous route to avoid detection. The fossil attracted considerable attention when it was exhibited at the Munich Fossil Show in Germany in 2011. While it was on display it was photographed by many people, one of whom was François Escuillié. Even though it was mostly held within the supporting rock matrix, the half of the skeleton that was visible was intriguing. The body was apparently that of a dromaeosaur, but with a head that was more bird-like and a snout-like that of a duck. Initial reactions to the photographs were that this was a faked fossil, an attempt to make a profit from a composite creation of a bird/reptile intermediate species. However, that this was genuine was demonstrated clearly when the fossil was scanned by high-energy X-rays generated by the European Synchrotron Radiation Facility at Grenoble in France. This produced images of the entire skeleton, including all that was embedded in the rock and previously unseen. The part which was hidden was a very precise fit with the visible section, adding to the belief that this was a genuine fossil. It also became apparent that the animal had wings, but in the form of penguin wings, of no use for flying, but distinctly wings. This strange and exotic fossil had come originally with an unknown provenance, which in itself causes problems for the process of dating the remains, but even more so for palaeoecology.

It was possible to determine the geographical origin of the specimen because the labelling indicated it had originated in Asia and the way in which the bones were preserved and the nature of the sediment in which it was set indicated that it was from the Gobi Desert. Ever more detailed analysis showed that it was possible to narrow it down further, with the sediment indicating an origin around Ukhaa Tolgod. By working out the geographical origin, it was possible to add a date to the fossil as well. It now looked as though *H. escuilliei* was from 71 to 75 MYA in the Cretaceous.

With all the available data, studying the fossil has become of great importance as it represents a very important evolutionary move. That it was originally considered to be a fake fossil can be understood from a sentence in Brownstein (2019) where he describes the animal as a 'clearly distinctive set of features that compose a superficially bizarre body plan'. When it was realised that this was a fossil from Mongolia that had been removed illegally, it was necessary for it to be returned. Great efforts were made by François Escuillié, so that it was eventually repatriated. He achieved this by negotiating with the dealer to buy the fossil, with the expressed intention of returning it to its country of origin.

Sometimes the misplaced fossils that turn out to have been smuggled have been moved on such a scale that it is impossible for a commercial transaction to take place without it becoming a very public exercise. This is what happened when an almost complete skeleton of *Tarbosaurus bataar* was put up for auction. This close relative of *T. rex* seems to have held the same ecological position in Asia that *T. rex* held in North America. *Tarbosaurus* fossils seem to be restricted to the Gobi Desert and surrounding areas of Mongolia and China, so any such skeleton would, by direct implication, have been removed illegally. In the case of the specimen of *Tarbosaurus* put up for sale by Heritage Auctions in New York in 2012, it was quite obvious that it had been smuggled and so became a subject of legal procedures to repatriate the skeleton. Of course, this had to be done using the correct legal procedure and as the question of jurisdiction of the law over the skeleton was one of power over an object the case had an unnerving element of satire, with the fossil being the defendant. The case which was heard in 2013 in New York was described as *United States* vs. *One Tarbosaurus bataar Skeleton*.

These particular examples indicate just how disruptive lack of provenance can be, but also, how without a demonstrable point of origin, how much the scientific value of a fossil can be reduced. To try and control the loss of data and maintain the cultural legacy that is associated with fossils, some countries have adopted a legal stance which states that fossils belong to the state and cannot be removed from the country without specific written permission from the correct government department. Two of the most prominent exponents of this control of assets are Mongolia and China. Both of these countries regard fossils found within their borders as belonging to the state, which sounds reasonable, but has some interesting and largely unexpected results. During the period from 2007 to 2010, it was reported that China had managed to repatriate approximately 5000 fossils that had been taken from unlicensed digs in China. This was reported in the English language China Daily, but no indication was given as to what constituted the recovered fossils. Under the pressure from a perception of ownership, new laws were introduced that came into force in 2011 giving the authorities the power to fine companies and individuals responsible for illegal removal of fossils (Stone 2010a).

An associated problem with smuggling, sometimes thought to have been exacerbated by legislation, is that of fakes. By making it illegal to export fossils from these areas, it has become possible for dealers to make private arrangements with unscrupulous collectors for illegally retrieved fossils, the origins of which are not closely investigated. Consequently, into this system, it is relatively easy to infiltrate unique, or special fossils, which are in fact fakes. It is also true that if the real thing is controlled and difficult to get hold of, why not sell a fake instead? For these unscrupulous collectors, it is of little significance that the fossil is fake as they are simply collectors, taking on trust what they are sold as genuine. If they were to reveal their material to a wider audience, they would risk losing the fossil and being open to legal action. But if these undiscovered fakes appear in museum collections to be studied, then all manner of confusion can occur. This has, it would appear, already taken place in some collections. It has been estimated that up to 80% of marine reptiles on display in Chinese museums have been altered, enhanced or combined from separate fossils to varying degrees (Stone 2010b).

The enhanced value of fossils to collectors who want objects from areas that do not permit their export has also affected the way in which illegal activity has taken place. In less strict times, skeletons would be removed from a dig in a more or less complete form, but now this is far more difficult to get away with. As an unexpected consequence, the smugglers are hacking and breaking up the fossil bearing strata in search of specifically saleable parts such as claws, skulls and teeth. This is in total disregard of the remaining fossilised bones and their importance for study. In many jurisdictions, the rules covering fossil material is much more lax. In most countries, the general rule is 'finders keepers' where the fossil belongs either to the owner of the land or to the person finding it, assuming it has been excavated with permission or it was on common land.

In recent years, the recognition that removing fossils illegally and publishing results based on research where the fossil is central to the work has some considerable ethical issues attached to it. It is for this reason during the early years of the twenty-first century that many journals have strict guidelines for publishing research on fossil material. These range from the fossil being in a collection which is accessible to outside researchers, to requiring written consent from the holder of the fossil and demonstrable legal provenance. Scientific societies, like the Geological Society of London, The Society of Vertebrate Palaeontology and the Palaeontological Association, have a range of requirements for publication in their journals, but also guidelines for field work. There is no doubt that high-value specimens that have been illegally collected and exported from their country of origin can exclude the local palaeontologists from research material, which may be exploited in the country where the fossil material ends up. This can unnaturally skew the research output in favour of wealthier nations who have gained access to these specimens illegally. It should be obvious that as a collaborative science palaeontology should really be working against the trend for smuggling and cultural insularity. This latter idea is a sensitive one which can limit access to material from the best researchers unless it, too, is seen as an unnecessary interference with scientific endeavour.

References

Beringer, J. (1726). *Lithographiae Wirceburgensis*. Würzburg University(trans. 1963 by M.E. Jahn and D.J. Woolf). USA: University of California Press.

Berry, W.B.N. and Gupta, V.J. (1967). Ordovician graptolites from Kashmir Himalayas. *Nature* 216: 1097.

Brownstein, C.D. (2019). *Halszkaraptor escuilliei* and the evolution of the paravian *bauplan*. *Scientific Reports* 9: 16455.

Cau, A., Beyrand, V., Voeten, D. et al. (2017). Synchotron scanning reveals amphibious ecomorphology in a new clade of bird-like dinosaurs. *Nature* 552: 395–399.

Cope, E.D. (1869a). On the reptilian orders *Pythonomorpha* and *Streptosauria*. *Proceedings of the Boston Society of Natural History* XII: 250–266.

Cope, E.D. (1869b). Synopsis of extinct Batrachia and Reptilia of North America part I. *Transactions of the American Philosophical Society* 14: 1–235.

Crichton, M. (1990). *Jurassic Park*. USA: Alfred A. Knopf.

Dalton, R. (2000). Feathers fly over Chinese fossil bird's legality and authenticity. *Nature* 403: 689–690.

Daumann, F. (2017). Wirtschaftliche Freiheit. *November* 12: 2017.

Dawson, C. and Woodward, A.S. (1913). On the discovery of a palaeolithic skull and mandible in a flint-bearing gravel overlying the Wealden (hasting beds) at Piltdown, Fletching (Sussex). *Quarterly Journal of the Geological Society* 69: 124–139.

Flannagan, J. (2019). Taste for 'ice ivory' mammoth tusks creates risk to Africa's elephants. *The Times* (22 August).

Grimaldi, D., Shedrinsky, A., Ross, A., and Baer, N. (1994). Forgeries of fossils in Amber: history, identification and case studies. *Curator* 37 (4): 251–274.

Gupta, V.J. and Denison, R.H. (1966). Devonian fishes from Kashmir, India. *Nature* 211: 177–178.

Hogarth, P.J. (1976). Ecological aspects of dragons. *Bulletin British Ecological Society* 7 (2): 2–5.

Hogarth, P.J. (1989). Ecological aspects of dragons. *Journal of Biological Education* 23 (2): 115–118.

Hutchinson, H.N. (1892). *Extinct Monsters*. New York, USA: D. Appleton and Company.

Jaffe, M. (2001). *The Gilded Dinosaur*. California, USA: Three Rivers Press.

Kreye, A. (2016). *Süddeutsche Zeitung* (8 July).

Kuhn, T. (1996). *Structure of Scientific Revolutions*, Thirde. USA: University of Chicago Press.

Leidy, J. (1870). Remarks on *Elasmosaurus*. *Proceedings of the Academy of Natural Sciences of Philadelphia* 22: 9–10.

Liston, J. (2016). Fossillegal: a symposium on ethics and palaeontology. *The Palaeontological Association Newsletter* 93: 27–31.

Marsh, O.C. (1890). Wrong end foremost. *New York Herald* (19 January).

Romer, A.S. (1933). *Man and the Vertebrates*. Chicago, USA: The University of Chicago Press.

Romer, A.S. (1959). *The Vertebrate Story*, 4e. Chicago, USA: The University of Chicago Press.

Ross, A.J. (1993). The Piltdown Fly. *Palaeontology Newsletter* Number 20, 16.

Sahni, M.R. and Gupta, V.J. (1964). Graptolites in the Indian sub-continent. *Nature* 201: 385–386.

Scott, S.P. and Rhines, R. (1975). Naming the Loch Ness monster. *Nature* 258: 466–468.

Shah, S.K. (2013). *Himalayan Fossil Fraud, The Palaeontological Society of India Special Publication*, vol. 4. India: University of Lucknow.

Sloan, C.P. (1999). Feathers for <T. rex>? New birdlike fossils are missing links in dinosaur evolution. *National Geographic* 196 (5): 98–107.

Stone, R. (2010a). China clamps down on illegal fossil trading. *Science* 329 (5998): 1453.

Stone, R. (2010b). Altering the past: China's faked fossil problem. *Science* 330 (6012): 1740–1741.

Waterston, D. (1913). The Piltdown mandible. *Nature* 92 (2298): 319.

Wheeler, H. (1939). *The Miracle of Life*. London: Odhams Press Limited.

Woodward, A.S. (1913). Note on the Piltdown man (*Eoanthropus dawsoni*). *Geological Magazine* 10 (10): 433–434.

Index

Page numbers in italics refer to illustrations.

Investigating Fossils: A History of Palaeontology, First Edition. Wilson J. Wall.
© 2021 John Wiley & Sons Ltd. Published 2021 by John Wiley & Sons Ltd.

CPSIA information can be obtained
at www.ICGtesting.com
Printed in the USA
LVHW080559270422
717290LV00015BA/444